CONVENTIONAL MILITARY STRATEGY IN THE THIRD NUCLEAR AGE

This volume delves into the way conventional deterrence operates between nuclear-armed states in the third nuclear age. Unlike the first and second ages the advent of this new age has witnessed greater strain on the principles of mutual vulnerability and survivability that may result in increased risks of advertent or inadvertent escalation and horizontal nuclear proliferation.

The book looks at the sum of three key simultaneous developments in the third nuclear age that merit attention. These include the emergence of asymmetric strategies, the introduction of unmanned platforms and the expansion of nuclear arsenals. The volume discusses how these concurrent developments might shape the practice of conventional deterrence and provides useful insights into conventional military dynamics, not just among the current nuclear dyads but also ones that may emerge in future. It seeks answers to several key issues in state security not limited to:

- What purpose and scope does the conventional military instrument have in a state's overall military strategy versus other nuclear-armed states?
- If mutual vulnerability and deterrence are the frameworks, why did the prospect of escalation appear in the first place?
- What are the trends – political, doctrinal, or technological – that augment or diminish conventional and nuclear interface?

With insights on military crises that have witnessed participation from nuclear-armed states like the United States, Russia, China, Pakistan, and India, this book will especially be of interest to scholars and researchers working in the areas of security and deterrence studies, defence and strategic studies, peace and conflict studies, and foreign policy. It will also appeal to policymakers, career bureaucrats, security and defence practitioners, and professionals working with think tanks and embassies.

Joy Mitra is a New Delhi–based analyst working on security issues spanning nuclear doctrine and posture development, conventional deterrence, and counter-terrorism with a focus on Southern Asia. He is a former visiting fellow at the Stimson Center's South Asia program and a former non-resident fellow at EastWest Institute. His commentary on these issues has appeared on The Wire, The Diplomat, Observer Research Foundation, South Asian Voices, Faultlines, and South Asia Intelligence Review, among others.

International Politics in the Age of Disruption

Series Editors:

International Politics in the Age of Disruption aims to bring together cross-sectoral expertise from across the globe, to inform and mediate academic and policy debates on contemporary global politics. It integrates sophisticated theoretical approaches in international politics and their application, with a policy-relevant focus for professional and academic audiences.

As multiple disruptions strain the post-Cold War global order, new strategic dilemmas for States have emerged in unpredictable ways to create challenges which few are equipped to grapple with effectively.

The volumes in this series explore a host of themes which include the changing global and regional balance of power; technological innovations, artificial intelligence, influence of Big Tech, and their impact on political societies; national security, surveillance, and warfare. The series also discusses non-traditional security threats and the post-pandemic world order; evolving socio-cultural trends such as populism; social media and political advocacy; challenges to economic globalisation; the emerging energy dynamics; and gender and politics, among others key issues in politics.

Techno-Geopolitics
U.S.-China Technology Competition and the Practice of Digital Statecraft
Pak Nung Wong

Understanding Security Role Evolution of US, China and India
Setting the Stage
Aditi Malhotra

Conventional Military Strategy in the Third Nuclear Age
Joy Mitra

For more information about this series, please visit:
https://www.routledge.com/International-Politics-in-the-Age-of-Disruption/book-series/IPAD

CONVENTIONAL MILITARY STRATEGY IN THE THIRD NUCLEAR AGE

Joy Mitra

Routledge
Taylor & Francis Group
LONDON AND NEW YORK

Cover image: Getty Images

First published 2023
by Routledge
4 Park Square, Milton Park, Abingdon, Oxon OX14 4RN

and by Routledge
605 Third Avenue, New York, NY 10158

Routledge is an imprint of the Taylor & Francis Group, an informa business

British Library Cataloguing-in-Publication Data
A catalogue record for this book is available from the British Library

ISBN: 978-1-032-41972-5 (hbk)
ISBN: 978-1-032-43169-7 (pbk)
ISBN: 978-1-003-36597-6 (ebk)

DOI: 10.4324/9781003365976

Typeset in Bembo
by Deanta Global Publishing Services, Chennai, India

CONTENTS

FIGURES

TABLES

ACKNOWLEDGMENTS

I am fortunate to have received support from a lot of people directly or indirectly while completing this project, and to all of them I owe a ton of gratitude.

To start with, I would like to mention two authors and their works, that I was deeply inspired from and were essential in shaping this text; "The Kill Chain: Defending America in the Future of High-Tech Warfare" by Chris Brose and "The Iraq War: Strategy, Tactics and Military Lessons" by Anthony, H. Cordesman. For someone who started with very little knowledge on these subjects their works were a godsend.

I would like to specially thank Dr Sreeram Chaulia, Prof. Mohsin Khan, and my former boss Dr Ajai Sahni. At different points in my career their support and encouragement have helped me carry on in my professional journey.

I am deeply thankful to Dr Amit Gupta, who was always very kind to take out time for discussions with me despite being very busy himself.

I cannot put in words how indebted I'm to Dr Harsh Pant and Dr Frank O'Donnell; they are some of the kindest and most generous persons in academia and I hope I can replicate some of that in my life and profession.

I am grateful to all my editors and reviewers for their direction; all the good work is a result of their suggestions, and all the mistakes are solely mine.

Finally, I'm indebted to my family and doctors for their endless support for me and for helping me get back on my feet after an injury.

INTRODUCTION

The advent of the nuclear era introduced two components in the security competition between states: one, it brought overwhelming destructive power and mutual vulnerability,[1] and second, it eroded the security dilemma by guaranteeing survival.[2] This rendered the concept of decisive victory outdated, defence was impossible, and deterrence took over as the preferred concept for viewing security between nuclear-armed states.[3] Scholars categorized this nuclear era into three ages.[4] The first nuclear age represented the Cold War–era bipolar competition between the United States and the Soviet Union characterized by arms races and destabilizing nuclear postures.[5] The second nuclear age saw horizontal nuclear proliferation and the emergence of regional nuclear powers,[6] who, unlike the Cold War participants, lacked the luxury of geography and the experience of arms control. Conventional conflict was not non-existent in these two ages: in the first it transpired on the territory of client states, and in the second, it transpired directly on the territory of nuclear-armed states. However, somewhere around the middle of the current decade the concept of a new third nuclear age began taking shape, marked by the potential emergence of several new nuclear powers, diffusion of destabilizing new technologies, and, most importantly, a new era of competition between great as well as new nuclear powers.[7]

The principles of mutual vulnerability and survivability that had hitherto ensured no major flare-ups between nuclear-armed states did not retain the same effect, leading to increased risks of horizontal nuclear proliferation and risks of advertent or inadvertent escalation.

Many different military crises witnessed participation from nuclear-armed states like the United States, Russia, China, Pakistan, and India, in this period,

DOI: 10.4324/9781003365976-1

in Syria, Ukraine, Kashmir, Doklam in the Ladakh region, Iran, South China Sea, and Taiwan. All of these military crises involved only conventional forces, yet the nuclear alarm bells were always ringing in the backdrop.

There are three key simultaneous developments in this phase of the third nuclear age that merit attention. These include the exercise of new kind of military strategies like "grey zone" or "hybrid warfare" which, though referred to as asymmetric strategies, are fundamentally dependent on being backed up by conventional forces; second, the, emergence of unmanned platforms along with other technologies, the sum of whom may impact the way forces are structured and battles fought; and, finally, the expansion of nuclear arsenals in both quantitative and qualitative terms to compensate for limitations in conventional forces, conventional deterrence or conventional strategies. This project essentially looks at the sum of these concurrent developments and how that might shape a new set of conditions for how conventional conflicts and military strategy might transpire especially in the future and the implications of that for the current set of nuclear armed states as well as the new ones.

In pursuit of that objective this project seeks answers to the following questions: first, if nuclear weapons are in play, mutual vulnerability is an established dictum, and deterrence is the framework, then why does the prospect of escalation appear in the first place? Second, what purpose and scope does the conventional military instrument have in a state's overall military strategy versus other nuclear-armed states? Third, what are the trends – political, doctrinal, or technological – that augment or diminish conventional and nuclear interface? Fourth, what are the factors that govern the shape, composition, and conduct of conventional forces? And, fifth, the overall thread that this project seeks to unravel in the process of answering the previous four is, how does the conventional military strategy balance between managing escalation and being an effective military instrument? This endeavour is actualized in the following four chapters of this book.

Chapter 1 deals with the scope and role afforded to the conventional military strategy and how advertent escalation responses are tied into conventional military doctrines and postures and the conceptual differences between such strategies that "instrumentalize" conventional capabilities versus nuclear coercion. This chapter deduces that there has been a shift from deterrence-based conventional strategies to conventional strategies based on the logic of compellence, where the conventional strategy is lent a cross-level framework of operation. Overall, it looks at the political intent that conventional strategies seek to serve and how advertent escalation responses tie into the strategy.

Chapter 2 looks at the new basket of technologies in the realm of data-processing and in the space domain to expound on the changes introduced in the kill chain and the manner in which the chain is operationalized. This chapter explains the shift from kill chain to kill web and covers the aspects of sensors, precision, range, and stealth. In the process it captures the cross-domain character of conventional operations and the risks of inadvertent escalation.

Chapter 3 looks at unmanned vehicles or drones, including autonomous drones, a key feature of the third nuclear age, and deals with the system-wide effects of the development of unmanned vehicles in terms of force structure, impact across the conflict spectrum, and trade-offs involved in their use and strategy. It touches upon the effects of the integration of unmanned autonomous vehicles in enabling concepts like mission commands and joint all-domain command and control and its implications on the conventional-nuclear interface.

Chapter 4 is the final chapter that evaluates the extent of segregation between conventional and nuclear strategies and domains. It determines the offence-defence balance at the different levels of the conventional conflict. This flows from the first chapter where offence dominance is located at the tactical level, which is then followed by defence dominance at the operational level. Multiple technologies and weapon systems – current, evolving, and emerging – are shown to operationalize an integrated offence-defence balance across these conventional levels of the conflict spectrum. However, the overall nature of the conventional-nuclear interface or nuclear ambiguity has a role to play here, and this space is again instrumentalized or used to address gaps in conventional deterrence.

Notes

1 Robert Jervis, "The Meaning of the Nuclear Revolution: Statecraft and the Prospect of Armageddon", Cornell University Press, 1989. pg. 22.
2 Francis J. Gavin, "Book Review Roundtable: The Meaning of the Nuclear Revolution 30 Years Later", Texas National Security Review, April 30, 2020, https://tnsr.org/roundtable/book-review-roundtable-the-meaning-of-the-nuclear-revolution-30-years-later/#_ftnref8
3 Ibid.
4 There is some debate about the conceptual difference between the second and third nuclear ages; some scholars argue that the current developments are an ongoing progress of the second nuclear age, while others see very specific trends that characterize the third nuclear age. For the former, see Hans Rühle & Michael Rühle, "Contours of a Third Nuclear Age", Real Clear Defense, July 18, 2017, https://www.realcleardefense.com/articles/2017/07/18/contours_of_a_third_nuclear_age_111832.html. And for latter see Nicholas L. Miller & Vipin Narang, "Is a New Nuclear Age Upon Us?", Foreign Affairs, April 16, 2020, https://www.foreignaffairs.com/articles/2019-12-30/new-nuclear-age-upon-us
5 Vipin Narang & Raghuveer Nidumolu, "The Third Nuclear Age", Carnegie India, 2019, https://carnegieindia.org/2019/08/23/third-nuclear-age-event-7170
6 Ibid.
7 Nicholas L. Miller & Vipin Narang, "Is a New Nuclear Age Upon Us?" Foreign Affairs, April 16, 2020, https://www.foreignaffairs.com/articles/2019-12-30/new-nuclear-age-upon-us

1

CONVENTIONAL MILITARY PURPOSE IN THE THIRD NUCLEAR AGE

Inter-state relationships that exhibit competition, particularly over any kind of long-term issue of contestation with a national security dimension to it, are often characterized by the use of military force or risk manipulation, to either maintain or force a change in status quo. When such military risk manipulation transpires between nuclear-armed states, they test a counter-intuitive logic of mutual preservation between adversaries to leverage what Thomas Schelling called "mutual deterrence".[1] To "deter" could be defined in one way as the ability to "prevent or discourage from acting by means of fear, doubt, or the like", or in another way as the ability "to turn aside or discourage through fear; hence, to prevent from action by fear of consequences".[2] The former, as described by Glenn Snyder, takes a cost-benefit approach to defining deterrence and calls it "risk calculus". Here the adversary is deterred if the expected cost of his actions outweighs the expected gains, and these are calculated on the basis of the net gain or net loss throughout the war rather than only the initial action outcomes.[3] The latter concept has more to do with "generation of fear"[4] and is generally associated with the use of nuclear weapons which carries the risk of mutually assured destruction (MAD) in nuclear dyads.

Neither form of deterrence implies that states stop competing, but that mutual hurt ensures that the competing states find a way to concert in political and military domains in a manner that prevents outbreak of hostilities or ensures that such an outbreak is essentially limited. The military dimension of this inter-state competition is further bifurcated into competition in the nuclear and conventional realms. The leveraging of mutual hurt is essential to both the nuclear and conventional realms of the conflict spectrum. Nuclear weapons states (NWSs)[5] must actualize capabilities that can deter the onset of conflict; further, they must ensure that if an outbreak of conventional hostilities does occur, it does not lead to nuclear annihilation.[6]

DOI: 10.4324/9781003365976-2

The political domain is not unrelated to these very simple, but essential, elements in the puzzle that materializes as the concept of deterrence between NWSs. One, the political domain sets the strategic objective of adhering to or maintaining the status quo or the use of force to change the status quo in favour of the state deploying force or any other instrument of coercion. Second, it decides the limit and the manner in which force, or another instrument of coercion, can be employed and the expendable resources needed to attain this strategic objective. The intersection of these limits and objectives sets the bracket in which the "logic of mutual hurt" can be leveraged effectively[7] in the conflict spectrum comprising conventional and nuclear realms. Military postures and strategies of NWSs, whether in the conventional or nuclear realms, are a function of this political intent and thereby of the conflict spectrum. There has, however, been a peculiar trend in conflicts where NWSs have been involved, particularly in the last two decades following the rise of new nuclear powers. This period, referred to as the Third Nuclear Age (TNA), has witnessed NWSs modernizing their nuclear weapon stockpiles qualitatively and quantitatively, accompanied by the emergence of those seeking independent nuclear arsenals.[8] A key characteristic attributed to the TNA has been hitherto unseen increased propensity for military risk between NWSs.[9] The concept of TNA itself does not expound on any systemic factors that may be responsible for this and the role of conventional forces and strategy within it. This chapter seeks to fill that void by drawing out the contours of conventional strategy space in the overall mix of offensive interests and nuclear weapons.

This text argues that a peacetime coercive intent can allocate proclaimed deterrence-based conventional strategies and forces a very distinct character in the overall conflict spectrum. This character is distinct from how it may have been originally operationalized as a deterrence by denial or deterrence by punishment strategy[10] within the conventional spectrum of the conflict. The interaction between the spectrum and the specific strategy in a certain realm of the conflict spectrum is referred to as the cross-level character of the conflict spectrum. This essentially implies that conventional strategy plays a dual role, one of compellence at a lower level (peacetime or sub-conventional) and another as a "firebreak" at the higher level (between conventional and nuclear levels); finally the scope for conventional strategy does not necessarily terminate at nuclear first use. This chapter starts by laying out a simple diagrammatic representation to visualize the concept of mutual hurt and its interaction with changing attributes of the conflict spectrum. It then moves to explain the concept of instrumentalized conventional strategy (ICS) and then expounds on the cross-level character or framework of operation for the conventional military strategy as observed across dyads. After contending with the question of whether conventional deterrence is failing and differentiating ICS from some alternative explanations, the text dives into the question of advertent escalation by use of tactical nuclear weapons (TNWs) and what implications that carries for the transition of the conflict from conventional to nuclear realm.

Figure 1.1 depicts how mutual hurt dynamically operates in the conflict spectrum between two NWSs, A1 and A2, where A1min-A1max and A2min-A2max are respectively the bracket of hurt for A1 and A2. A2min-A1max represents the intersection area in which mutual hurt operates between the two NWSs. A stronger and more credible NWS will have a smaller area of hurt, and its conventional or nuclear capabilities and posture will decide whether this area of hurt has a greater or lesser overlap with the conventional or nuclear realm. Here NWS A2 is a conventionally stronger state than NWS A1; this firmly establishes that A2min-SC > A1min-SC, where SC represents the lower end of the conventional spectrum. SC-CN max represents the conventional spectrum where SC-CN min is the purely conventional spectrum and CN min-CN max is the "overlap" space where the distinction between conventional and nuclear realm is blurred, though the primary character is still conventional. The point CN max represents the "transition" point between conventional-nuclear realms.

This bracket of hurt works on the logic that a state is more risk-prone outside its own bracket of hurt and risk-averse inside it. The equation between A1min-A1max and A2min-A2max space is indeterminant, and all three relations A1min-A1max>A2min-A2max, A1min-A1max=A2min-A2max, and A1min-A1max<A2min-A2max are possible. This possibility exists as NWS A1 is more inclined to escalate to nuclear warfighting decreasing its area of hurt over the nuclear realm, while A2 may be less inclined to fight in the nuclear spectrum increasing its area of hurt over the nuclear realm. Depending on peacetime, crisis, or any other unique point in the conflict and thereby the deterrence relationship at that point of time the bracket arrangement can represent variable tolerance for conflict along the spectrum.

Therefore, if the A2min-A1max intersection space is large, that effectively implies greater bilateral deterrence and conversely if A2min-A1max is small, a lower state of bilateral deterrence. The A2min-A1max space can be small on two

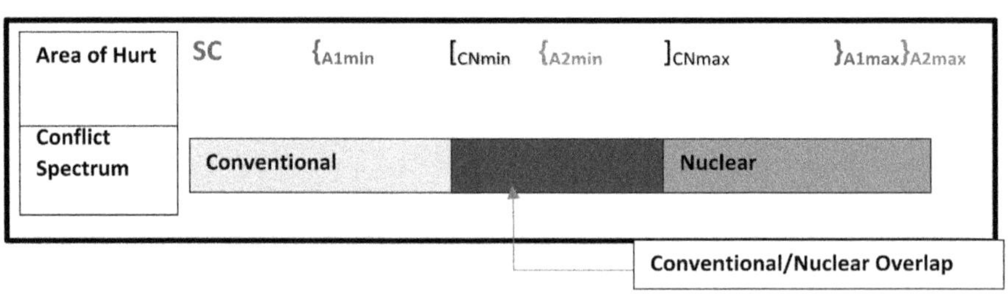

FIGURE 1.1 Bracket of mutual hurt. Note: Assumption here is that the mutual hurt bracket is the end intersection of each actor's perception of where its own hurt bracket lies along with its adversary's over the conflict spectrum. Different combinations of weak and strong conventional adversaries can be arrived at by adjusting these brackets and the spectrum over which they intersect where they can inflict damage through their capabilities on the adversary.

counts: one if the intersection area is less, or the other possibility that both states have significantly smaller areas of hurt despite having perfect intersection. This latter reasoning is improbable and unlikely, because it will lead both states to be risk-prone in higher levels of both nuclear and conventional realms with perfect deterrence between the two.

Figure 1.2 shows a peacetime arrangement between NWSs A1 and A2. In peacetime A2min-A1max space would sit astride the point CN max in the conflict spectrum to cover higher levels of the conventional ladder and the nuclear level. The upper bounds of both states A1 max and A2 max lie at almost the extreme right end of the nuclear spectrum, exhibiting working nuclear deterrence but different tolerances for lower- or mid-level conventional provocations.

One could argue for an alternative argument that a smaller A2min-A1max space tending towards nil intersection could afford higher bilateral deterrence because that would expand the overall area of hurt in the dyad. This, however, is unlikely on two counts: first this would place one NWS, say A1's, bracket of hurt completely outside NWS A2's bracket of hurt (see Figure 1.3). Such mutual exclusivity is improbable as it would imply weaker A1 is risk-averse to low-level conventional provocations but is risk-prone to tolerating higher-level conventional or nuclear-level actions. In peacetime as well as in a crisis such a unique spread of the bracket of hurt will disobey the fundamentals of nuclear deterrence. Second, this is a game of imperfect information where the adversary can perfectly see his own bracket of hurt but has incomplete information about his

FIGURE 1.2 Peacetime arrangement.

FIGURE 1.3 Flawed peacetime arrangement.

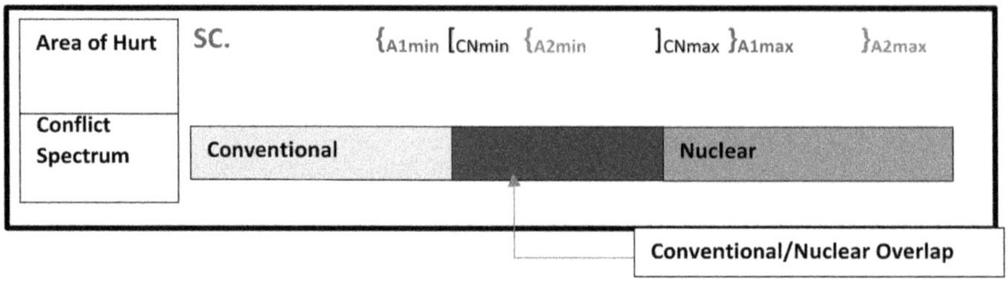

FIGURE 1.4 Crisis arrangement.

adversary's bracket of hurt. Therefore, it is not the union of the brackets of hurt but rather the intersection that defines the space in which mutual hurt operates.

In Figure 1.2, SC-A1 is the space where actor A1 operates below A2's threshold of conventional military conflict; however, it fears conventional escalation by A2 in the A1min-A2min space. During a crisis (Figure 1.4) both A1 min and A2 min shift right, opening up more conventional levels, including the ones in the overlap region. A2 being the stronger conventional power opens up more conventional-level options, and hence A1min-CNmax is greater than A2min-CNmax. At a certain point in the crisis, A1 max and A2 max shift leftward, but the A1 max shift is greater as A1 intends to compensate for relative conventional weakness against A2 with risk manipulation of its tolerance for a nuclear conflict. For NWS A1 the incentive is always to avoid higher-degree conventional confrontation with A2, thus signalling its willingness to shift the conflict in the nuclear spectrum. This leads A1 to decrease its bracket of hurt over the higher-degree responses particularly in the nuclear spectrum such that A1max-CNmax < A2max-CNmin and A1max-CNmax space approaches zero. This results in a reduction of the A2min-A1max space of mutual hurt, thereby reducing effective deterrence between the two states.

The three spaces relevant to understanding conventional military strategy here are the spaces SC-A1min or SC-A2 min where A1 can afford to challenge A2 using its conventional forces at the lower rungs of the conventional spectrum or vice versa; second, CNmin-CNmax the conventional-nuclear "overlap" area; and finally, the conventional-nuclear "transition" point.

The CNmin-CNmax overlap area is essentially the space where "entanglement" of conventional and nuclear systems presents the most potential for inadvertent escalation. Co-location nuclear and conventional forces, weapon systems, or facilities that present targeting ambiguity for the adversary or reduce the distinguishability that forces, particularly dual-capable platforms, add to this space. Proliferation of more capable Ballistic Missile Defence (BMD) systems; reliance of conventional capabilities on space-based assets that are shared with nuclear or strategic forces for situational awareness, command, and control; and finally highly precise conventional strike options that mirror the impact of nuclear weapon also form a part of this space. Under ideal conditions if conventional counter-force strategies were to be realized to their full potential, that effectively will merge the nuclear and conventional spectrum completely.[11] The investigation and study of this space and corresponding factors of inadvertent escalation will be discussed in later chapters. More iterations of NWS dyads can be arrived at by changing the mutual-hurt bracket and the character of the conflict spectrum bar.

The dyad explained above exhibits attributes that are general to most, if not all, NWSs and highlights the challenges NWSs face in formulating their conventional military strategy and the role of their conventional deterrent (and their nuclear deterrent) against other NWSs. Conventional military strategy is effectively a function of the conflict spectrum continuum in the sense that the spectrum and its intersection with the mutual area of hurt shape the space for limited conventional conflict. On one end the political domain sets the threshold for what is tolerable during peacetime, and thereby "what is not" tolerable to trigger a conventional response. This sets the bar for the lower end of the conventional conflict space and trickles down in the form of the doctrines and concepts that militaries utilize to operationalize their conventional strategy against the adversary. At the other end it has the task of achieving these politico-military objectives whilst avoiding strategic nuclear escalation.

As the dyad above shows NWSs can design nuclear "first-use" doctrines to account for weak conventional capability effectively pulling down the upper bound (CN max) for conventional military responses. When combined with the changes in conventional strategy objectives at the lower bound, they lead to a systemic trend of instability and tolerance for escalation across the conflict spectrum. This change at the lower bound is brought about by the existence of peacetime revisionist tendencies that are either territorial or behavioural in nature, a key feature of the TNA.[12] A natural outcome of this peacetime coercion is that conventional strategies, though built around the framework of deterrence and dissuasion in their own conflict spectrum that is within a conventional engagement of military forces, tend to acquire an overall offensive or compellent intent.[13] Coercion can take both deterrence and compellence forms. However, compellence, unlike deterrence, is not passive or preventive, but is rather pre-emptive and forces a "desired action" on the adversary.[14] Like brute force, it has elements of unilateral action but leaves a choice for the adversary to comply (or not), whereas in case of brute force action a new status quo is established regardless of the other actor's consent.[15] TNA evinces a trend of compellent objectives that are instrumentalized by conventional forces and strategies, which may be operating within the framework of deterrence by denial, punishment, or a combination of the two in its own conventional spectrum. This gives conventional strategy a cross-level coercive framework that is explained in the next section.

Deterrence to compellence: doctrines, postures, and instrumentalized conventional strategy

This section will explore SC-A1min/SC-A2min space, the trends that have changed to influence the political intent and the consequential impact on the military domain, resulting in a very different purpose and role for conventional military strategy in the TNA. The most important of these developments is the continuous increase in peacetime coercion throughout the second and third nuclear ages. This form of coercion does not preclude the use of state forces or special forces/operatives; therefore, below the military threshold characterization is not correct. In fact, all asymmetric strategies, whether they are "hybrid warfare" in Ukraine, "grey-zone warfare" in South China Sea, or "proxy warfare" in Jammu and Kashmir, are essentially deeply integrated and fundamentally dependent upon conventional military operations that enable them. The term used in this text, "instrumentalized conventional strategy" (ICS), therefore denotes actions that necessarily include some elements of conventional state forces or state-led violent action over an appreciable time interval, are calibrated in their use, and have a strategic intent of altering the status quo. In addition, the gains made through ICS are mutually reinforcing with the conventional military posture which is gradually fortified and in turn enables further such coercion. This also means that this form of coercion is essentially not "sub-conventional" in character since conventional forces are heavily invested and utilized in backing such coercion.

This peacetime coercion between NWSs is a feature that currently exists in both extended and direct deterrence dyads. The premise of all such coercion is an alteration of the status quo or "pre-emption" of military conflict by using non-military and conventional military means, sometimes with deniability (or not) for inflicting a military outcome. This emphasis on the alteration of the status quo is a new introduction to deterrence dynamics between nuclear dyads and a trend that has proliferated in the TNA. The military outcome is imposed slowly over time and is grounded in a strategy of compellence or more specifically "fait accompli".[16] Here a threatening action is taken in a gradual limited manner to establish a new status quo, and the choice is then left to the adversary to accept or have the onus of counter-escalation. Forceful action is central to compellence, including fait accompli cases where military strategy and posture are important factors driving the fait accompli. The two important elements of this strategy are establishing a new status quo and leaving the onus of escalation on the adversary.

These elements are not a complement to the military capabilities in the conventional spectrum; in fact, conventional strategies are built around them.[17] As a ramification conventional strategy acquires a role that is supportive towards the objective of coercion. The term "conventional deterrence" can be slightly misleading in that sense as, though the operational framework for conventional capabilities may be "deterrence", they are acquired and essentially instrumentalized by peacetime coercive strategies which are inherently based on the logic of "compellence" (see Figure 1.5). In that sense, conventional strategy and, by

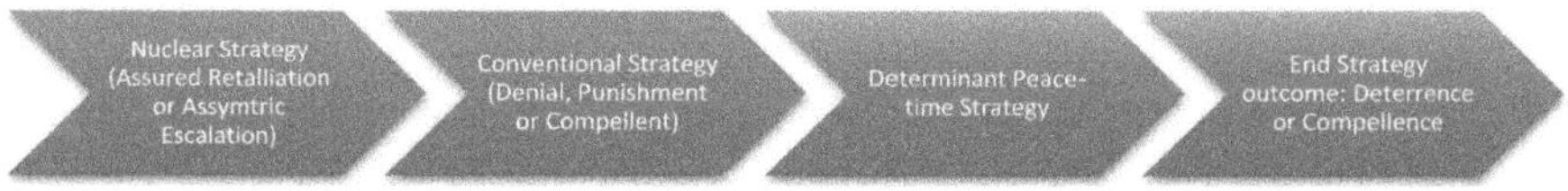

FIGURE 1.5 Cross-level instrumentalized strategy.

extension, in some cases, even nuclear strategy in effect serve an overall purpose of compellence.[18] This is not a surprising finding given the fact that conflict spectrum over time has become a continuum where there are no clear boundaries between sub-conventional, conventional, and nuclear levels. In a way, the agency factor for whether a military strategy operates within the deterrence or compellence framework lies at the lowest level of the continuum.[19]

States using ICS may have a conventional strategy that seeks "punishment" or "denial" in order to deter military actions against their newly established objectives. Though these strategies are built around the framework of deterrence, they, however, serve compellent objectives. As a ramification, there is a proliferation of offensive conventional strategies and concepts, heavy on the object of deterrence in theory but that are in fact embedded in overall compellent strategies (see Tables 1.1 and 1.2).

If this leads to a failure of deterrence or failure in the realization of the objective of deterrence, then this is bound to further impact conventional military strategies of NWS that have hitherto largely been based on the framework of deterrence.[20] Repeated failure in establishing deterrence could lead to more compellent behaviour, thus increasing the overall risk-prone activity between nuclear-armed states, a key characteristic of TNA.

This is not to say that conventional deterrence and strategies do not have the goal of resisting the outbreak of nuclear hostilities, but that this goal is importantly secondary in determining the character and purpose of such strategies, as opposed to the sequentially first and dominant aspect of resisting or instrumentalizing peacetime coercion.[21] In some cases conventional strategy in pursuit or resistance of a compellent objective forces the adversary's nuclear strategy calculations in a manner that they integrate into or substitute for conventional deterrence or the lack of it. This leads to an end-state where both adversaries in effect perceive the intent of conventional military instruments as offensive, irrespective of the deterrent or defensive intent within the conventional level in the conflict spectrum. This cross-linkage within the overall conflict spectrum affords conventional strategy a cross-level,[22] rather than simply a cross-domain,[23] character because it is instrumentalized to achieve compellent objectives (see Figure 1.6)

The next section will uncover evidence for instrumentalized conventional strategy (ICS) that serves peacetime coercion in four nuclear dyads of interest and highlight the objective, modus operandi of compellence, and the need for mutually reinforcing nature of the gains with the conventional posture of the

TABLE 1.1 Summary of doctrine-posture attributes of NWSs in conventional and nuclear domains

Strategy/Posture Attribute	*Conventional Doctrine (Offensive/Defensive/Deterrent)*[a] *and WarFighting Concepts*	*Nuclear Doctrine*	*Nuclear Posture*
China	Offensive[b] Strike-First[c] Winning Informationized Limited/Local War[d e f] on the periphery	No first use[g h] Limited ambiguity in case the nuclear facilities are attacked with conventional strikes[i]	Assured Retaliation[j] Launch on Warning (LoW)[k l]
India	Offensive[m] "Cold Start doctrine" for limited, calibrated and swift[n] conventional operations[o] Pro-active strategy against sub-conventional threats ("Surgical Strikes") [p q] Conflict prevention through deterrence or use of force[r]	No first use[s] Massive Retaliation[t]	Assured Retaliation[u] Launch on Hit (LoH) or Launch on Warning (LoW)[v w x]
Pakistan	Deterrent[y] New Concept of War Fighting (NCFW)[z aa] "Quid-Pro-Quo-Plus"[ab]	First use with ambiguous territorial, economic, military and stability thresholds[ac] "Full-Spectrum Deterrence"[ad]	Asymmetric Escalation[ae] De-mated and non-deployed posture[af]
Russia	Offensive[ag] Prevention and Pre-emption of Military Conflict[ah] "Non-Nuclear Deterrence"[ai]	Nuclear first use against an existential threat (conventional or nuclear) [aj al al] Arrival of reliable data on launch of ballistic missiles targeting Russian territory; actual use of nuclear weapons or any other WMD against Russian federation or its allies[am] Attack (nuclear, conventional, cyber, electronic) against critical government or military centres undermining nuclear forces response[an]	Asymmetric Escalation[ao] [ap] and "Reciprocal Counter-Strike" [aq ar] Launch on Warning (LoW)[as]

United States	Deterrent[at] "tailored deterrence"[au]		"Nuclear first use in extreme circumstances including against non-nuclear strategic attacks"[av] Assurance: "Including attacks against US, allied or partner civilian populations or infrastructure, or against US or allied nuclear infrastructure"[aw] "flexible and limited response"[ax]	Asymmetric Escalation and Reciprocal counterstrike[ay] Launch on Warning (LoW)[az]

[a] For military doctrine classification, see Barry R. Posen. "The Sources of Military Doctrine: France, Britain, and Germany between the World Wars". Cornell University Press, 1984, www.jstor.org/stable/10.7591/j.ctt1287fp3.

[b] "Analysis: China's Evolving Military Doctrine after the Cold War", SETA, February 25, 2020, https://www.setav.org/en/analysis-chinas-evolving-military-doctrine-after-the-cold-war/

[c] Ibid.

[d] Taylor Fravel. (2018) Shifts in warfare and party unity: Explaining China's changes in military strategy, International Security, 42:3, 37-–83, pg. 50, 51.

[e] IDSA Fellows Discussion. "The Evolution of Chinese Military Doctrine". Manohar Parrikar Institute for Defence Studies and Analyses, 2016, https://idsa.in/event/TheEvolutionofChinasMilitaryDoctrines

[f] M. Taylor Fravel. "Active Defense: China's Military Strategy since 1949". Princeton University Press, 2019. pg. 218–219.

[g] Fiona S. Cunningham & M. Taylor Fravel, "Assuring Assured Retaliation: China's Nuclear Posture and U.S.-China Strategic Stability", International Security, 2020, https://www.mitpressjournals.org/doi/abs/10.1162/ISEC_a_00215

[h] "Full Text: China's National Defense in the New Era", Www.gov.cn, 2019, http://english.www.gov.cn/archive/whitepaper/201907/24/content_WS5d3941ddc6d08408f502283d.html

[i] Fiona S. Cunningham 7 M. Taylor Fravel, "Assuring Assured Retaliation: China's Nuclear Posture and U.S.-China Strategic Stability", International Security, 2020, https://www.mitpressjournals.org/doi/abs/10.1162/ISEC_a_00215

[j] "Full Text: China's National Defense in the New Era", Www.gov.cn, 2019, http://english.www.gov.cn/archive/whitepaper/201907/24/content_WS5d3941ddc6d08408f502283d.html

[k] See "Fear of False Negatives: AI and China's Nuclear Posture - Bulletin of the Atomic Scientists", Bulletin of the Atomic Scientists, April 25, 2018, https://thebulletin.org/2018/04/fear-of-false-negatives-ai-and-chinas-nuclear-posture/

[l] China's Military Strategy, "《中国的军事战略》白皮书（全文）", Scio.gov.cn, 2015, http://www.scio.gov.cn/zfbps/ndhf/2015/Document/1435161/1435161.htm

(Continued)

TABLE 1.1 Continued

[m] Author's deduction.

[n] "India's Limited War Doctrine: The Structural Factor", Manohar Parrikar Institute for Defence Studies and Analyses , 2016, https://idsa.in/monograph/IndiasLimitedWarDoctrine_aahmed

[o] "'We Will Cross Again'", India Today, January 6, 2017, https://www.indiatoday.in/magazine/interview/story/20170116-lt-general-bipin-rawat-surgical-strikes-indian-army-985527-2017-01-04

[p] "Joint Doctrine Indian Armed Forces", https://www.ids.nic.in/IDSAdmin/upload_images/doctrine/JointDoctrineIndianArmedForces2017.pdf

[q] Land Warfare Doctrine, Indian Army, 2018.

[r] Indian Military, Joint Doctrine Indian Armed Forces, https://www.ids.nic.in/IDSAdmin/upload_images/doctrine/JointDoctrineIndianArmedForces2017.pdf. pg. 14.

[s] "The Cabinet Committee on Security Reviews Operationalization of India's Nuclear Doctrine", Ministry of External Affairs India, 2013, https://mea.gov.in/press-releases.htm?dtl/20131/The_Cabinet_Committee_on_Security_Reviews_perationalization_of_Indias_Nuclear_Doctrine+Report+of+National+Security+Advisory+Board+on+Indian+Nuclear+Doctrine

[t] Ibid.

[u] "The Cabinet Committee on Security Reviews Operationalization of India's Nuclear Doctrine", Ministry of External Affairs India, 2013, https://mea.gov.in/press-releases.htm?dtl/20131/The_Cabinet_Committee_on_Security_Reviews_perationalization_of_Indias_Nuclear_Doctrine+Report+of+National+Security+Advisory+Board+on+Indian+Nuclear+Doctrine

[v] Joy Mitra, "SAV Review: When Push Comes to Shove, Will India's NFU Stand?", South Asian Voices, January 9, 2019, https://southasianvoices.org/sav-review-when-push-comes-shove-will-india-nfu-stand/

[w] Christopher Clary & Vipin Narang (Winter 2018/19) India's counterforce temptations: Strategic dilemmas, doctrine, and capabilities, International Security, 43:3, 7–52, DOI: 10.1162/ISEC_a_00340

[x] Frank O'Donnell (2017) Reconsidering minimum deterrence in South Asia: Indian responses to Pakistan's tactical nuclear weapons, Contemporary Security Policy, 38:1, 78–101, DOI: 10.1080/13523260.2016.1236470, 16.

[y] "Keynote Address and Discussion Session with Lieutenant General (Retd) Khalid Kidwai", International Institute of Strategic Studies, 2020, https://www.iiss.org/events/2020/02/7th-iiss-and-ciss-south-asian-strategic-stability-workshop

[z] "Countering Cold Start: Military to Adopt New War Concept", The Express Tribune, June 4, 2013, https://tribune.com.pk/story/558604/countering-cold-start-military-to-adopt-new-war-concept/

[aa] Meenakshi Sood, "Pakistan's (Non-Nuclear) Plan to Counter 'Cold Start'", The Diplomat, March 25, 2017, https://thediplomat.com/2017/03/pakistans-non-nuclear-plan-to-counter-cold-start/

[ab] "Keynote Address and Discussion Session with Lieutenant General (Retd) Khalid Kidwai", International Institute of Strategic Studies, 2020, https://www.iiss.org/events/2020/02/7th-iiss-and-ciss-south-asian-strategic-stability-workshop

[ac] "Report on Nuclear Safety, Nuclear Stability and Nuclear Strategy in Pakistan", Pugwash Conferences on Science and World Affairs, January 14, 2002, https://pugwash.org/2002/01/14/report-on-nuclear-safety-nuclear-stability-and-nuclear-strategy-in-pakistan/

[ad] "No PR-133/2013-ISPR", Inter Services Public Relations, September 5, 2020, https://www.ispr.gov.pk/press-release-detail.php?id=2361

[ae] Vipin Narang (Winter 2009/10) Posturing for peace? Pakistan's nuclear postures and South Asian stability, Quarterly Journal: International Security, 3:3, 38–78, pg. 44.

[af] Sadia Tasleem, "Pakistan's Nuclear Use Doctrine", Carnegie Endowment for International Peace, 2016, https://carnegieendowment.org/2016/06/30/pakistan-s-nuclear-use-doctrine-pub-63913

[ag] Author's deduction.

[ah] "THE MILITARY DOCTRINE OF THE RUSSIAN FEDERATION", **Посольство России в Великобритании**, 2020, https://rusemb.org.uk/press/2029

[ai] Ibid.

[aj] Ibid.

[ak] Mike Eckel, "Putin Says Russia Has Ruled Out Launching Preemptive Nuclear Strikes", Radio Free Europe / Radio Liberty, October 19, 2018, https://www.rferl.org/a/putin-says-russia-has-ruled-out-launching-preemptive-nuclear-strikes-only-use-defensively/29552055.html

[al] **РФ готова применять ядерное оружие в локальных конфликтах**, "**РФ Готова Применять Ядерное Оружие в Локальных Конфликтах**", **РБК**, October 14, 2009, https://www.rbc.ru/politics/14/10/2009/5703d6e19a7947733180bb63

[am] "Basic Principles of State Policy of the Russian Federation on Nuclear Deterrence", Ministry of Foreign Affairs, Russia, June 2, 2020, https://www.mid.ru/en/web/guest/foreign_policy/international_safety/disarmament/-/asset_publisher/rp0fiUBmANaH/content/id/4152094#:~:text=State%20policy%20on%20Nuclear%20Deterrence%20is%20defensive%20by%20nature%2C%20it,adversary%20from%20aggression%20against%20the

[an] Ibid.

[ao] "THE MILITARY DOCTRINE OF THE RUSSIAN FEDERATION", **Посольство России в Великобритании**, 2020, https://rusemb.org.uk/press/2029 and

[ap] For the arguments against lowered Russian nuclear threshold, see Kristin Ven Bruusgaard, "The Myth of Russia's Lowered Nuclear Threshold", War on the Rocks, September 22, 2017, https://warontherocks.com/2017/09/the-myth-of-russias-lowered-nuclear-threshold/ and Franz-Stefan Gady, "Interview: Dmitry Stefanovich on Russia's Nuclear Forces and Doctrine", The Diplomat, November 6, 2019, https://thediplomat.com/2019/11/interview-dmitry-stefanovich-on-russias-nuclear-forces-and-doctrine/

[aq] "Meeting of the Valdai International Discussion Club", President of Russia, October 18, 2018, http://en.kremlin.ru/events/president/news/58848

[ar] Franz-Stefan Gady, "Interview: Dmitry Stefanovich on Russia's Nuclear Forces and Doctrine", The Diplomat, November 6, 2019, https://thediplomat.com/2019/11/interview-dmitry-stefanovich-on-russias-nuclear-forces-and-doctrine/

(*Continued*)

TABLE 1.1 Continued

[as] "Putin Says Russia Would Only Use Its Nuclear Weapons in Retaliation", PBS NewsHour, October 18, 2018, https://www.pbs.org/newshour/world/putin-says-russia-would-only-use-its-nuclear-weapons-in-retaliation

[at] Author's deduction.

[au] "Special Report: Nuclear Posture Review - 2018", U.S. Department of Defense, 2018, https://dod.defense.gov/News/SpecialReports/2018NuclearPostureReview.aspx.

[av] Ibid, pg. 21.

[aw] Ibid, pg. 21.

[ax] Ibid, pg. 23.

[ay] Ibid, pg. 22.

[az] "Frequently Asked Questions about Taking Nuclear Weapons Off Hair-Trigger Alert", Union of Concerned Scientists, https://www.ucsusa.org/sites/default/files/attach/2015/01/Hair-Trigger%20FAQ.pdf

TABLE 1.2 Summary of attributes of coercion across dyads

Strategy/Posture Attributes	*Nature of Peacetime Coercion*		*Conventional Deterrence Form*[a]	
Dyad (A1/A2)	A1	A2	A1	A2
US (NATO)/Russia	Objective: Territorial/Behavioural & Revisionist Economic Sanctions[b] Forward deployment of missile defence capable systems in the European theatre[c] Eastward expansion of NATO[d]	Objective: Territorial & Revisionist Cyber[e] and information warfare including interference in domestic elections[f] Use of ambiguous tactical operations for strategic purpose	Mix of Deterrence by Denial and Punishment	Deterrence by Punishment
US/China	Objective: Territorial/Behavioural & Status quoist Freedom of Navigation Operations (FONPs)	Objective: Territorial & Revisionist Influence operations and implied military threats[g] Maritime Militia	Mix of Deterrence by Denial and Punishment	Deterrence by Punishment
India/China	Objective: Territorial and status quoist Diplomatic, Legal along with threat of military action	Objective: Territorial and revisionist Lawfare, military construction and aggressive patrolling for Salami Slicing	Deterrence by Punishment	Deterrence by Punishment[h]
India/Pakistan	Objective: Behavioural revisionist[i] and territorial status quoist Diplomatic, Economic and Legal[j] along with threat of limited conventional military action	Objective: Territorial/Behavioural and Revisionist Asymmetric Warfare using militant actors[k] Economic warfare using counterfeit currency[l]	Deterrence by Punishment	Mix of Deterrence by Denial and Punishment

(*Continued*)

TABLE 1.1 Continued

[a] The conventional deterrence stances take the form of denial or punishment. Though an "absolute sharp distinction" between the two is hard to make, Yogesh and Anit highlight three key distinctions which have been used to classify these deterrence positions: "These are, one, "dominant function" of the deterrent military action whether the idea is to simply stall the enemy or to raise costs for its action. Second is the theatre of deterrent military action if it is on one's own territory or on the enemy's turf. Thirdly a pure denial strategy on the part of the defender may communicate to the adversary that hostilities will be limited to its preferred area of action, whereas a punishment strategy will entail horizontal expansion of the conflict." This classification has also been used here. See Yogesh Joshi & Anit Mukherjee (2019) From denial to punishment: The security dilemma and changes in India's military strategy towards China, Asian Security, 15:1, 25–43 and Glenn H. Snyder. "Deterrence by Denial and Punishment". Research Monograph No. 1, Woodrow Wilson School of Public and International Affairs, Princeton University, January 2, 1959. pg. 1.

[b] "U.S. Sanctions against Russia: What You Need to Know", Center For Strategic and International Studies, 2018, https://www.csis.org/analysis/us-sanctions-against-russia-what-you-need-know

[c] National Security Strategy of the United States of America, DECEMBER 2017, pg. 48.

[d] "Threat or Threatened? Russia in the Era of NATO Expansion", Australian Institute of International Affairs, 2018, http://www.internationalaffairs.org.au/australianoutlook/threat-or-threatened-russian-foreign-policy-in-the-era-of-nato-expansion/

[e] Andy Greenberg, "The Untold Story of NotPetya, the Most Devastating Cyberattack in History", Wired, August 22, 2018, https://www.wired.com/story/notpetya-cyberattack-ukraine-russia-code-crashed-the-world/

[f] "Russia Engaging in 'information Warfare' Ahead of 2020 Election, FBI Chief Warns", The Guardian, February 6, 2020, https://www.theguardian.com/us-news/2020/feb/06/russia-hacking-information-warfare-election-2020

[g] National Security Strategy of the United States of America, DECEMBER 2017, pg. 46.

[h] Zhang Tao, "Considerations for Replacing Military Area Commands with Theater Commands", China Military Online, 2016, http://eng.chinamil.com.cn/news-channels/2016-02/03/content_6888462.htm

[i] Frank O'Donnell, "Reconsidering Minimum Deterrence in South Asia: Indian Responses to Pakistan's Tactical Nuclear Weapons", Contemporary Security Policy, 2016, DOI: 10.1080/13523260.2016.1236470, 11.

[j] "Why India's Pakistan Policy Needs a Rethink", The Economic Times, December 11, 2019, https://economictimes.indiatimes.com/news/politics-and-nation/why-indias-pakistan-policy-needs-a-rethinking/articleshow/72480835.cms?from=mdr

[k] Christine Fair, Frederic Grare & Ashley J Tellis, "Fighting to the End: The Pakistan Army's Way of War", Carnegie Endowment for International Peace Event, 2014, https://carnegieendowment.org/2014/09/04/fighting-to-end-pakistan-army-s-way-of-war-event-4536.

[l] "Pakistan Officially Printing Fake Indian Currency Notes", India Today, January 22, 2014, https://www.indiatoday.in/india/north/story/nia-fake-indian-currency-notes-pakistan-216536-2013-11-06

	Level or Realms		
	Peacetime/Sub conventional Level	Conventional Level	Nuclear Level
Domains		Space	
		Cyber	
		Electronic	
		Air/Land/Sea	

FIGURE 1.6 Level versus domain.

state presenting the fait accompli. This will incorporate a discussion of the proclaimed doctrines, posture attributes, and warfighting concepts (see Table 1.1) that underpin ICS.

In the United States and North Atlantic Treaty Organization (NATO)-Russia dyad, after 2000 a concept of "Russian Strategic Deterrence" started taking shape which was formalized into the doctrine by 2010.[24] This concept endeavoured to integrate conventional, nuclear, and non-military capabilities in situations of peace, conflict, and war.[25] The nuclear threat[26] was certainly involved but not pivotal to it, and it existed only to deter large-scale aggression.[27] It combined diplomatic, economic, and political instruments, information warfare, cyber and special operation forces along with other kinetic and non-kinetic military capabilities. In the final version of 2014 (see Table 1.1) this got crystallized as the concept of non-nuclear deterrence essentially with the objective of "prevention of military conflict". General Valery Gerasimov described this as an "active defence strategy" that "provides for a set of measures to proactively neutralize threats to the security of the state".[28] He adds, "We must act quickly so as to pre-empt the enemy with our preventive measures, promptly identify his vulnerabilities, and create threats of unacceptable damage to it. This ensures that the strategic initiative is captured and maintained".[29] This ostensibly was the Russian approach in Ukraine, where NATO's perceived offensive interest[30] led it to orchestrate a "controlled chaos" with tactical but non-attributable action.[31]

This was backed with the concentration of conventional forces, conventional long-range precision strikes and high-tech standoff capability,[32] and deliberate nuclear messaging.[33] The nuclear messaging wasn't pivotal to the form of coercion effected, as in reality NATO faced a conventional capability gap versus Russia in Crimea to the extent that Baltics would fall to Russia within days.[34] Russian could simply employ conventional artillery, airpower, submarines, and long-range missiles to take out the NATO forces in the rear.[35] In other words the "threats of unacceptable damage" created were conventional in nature, not hybrid or irregular or nuclear. A similar approach was taken in Georgia, where Russian forces grabbed territory in South Ossetia.[36] Conventional compellence was exercised with a military deployment posture[37] such that the onus of conventional escalation rested on NATO.[38] Conventional forces were directly utilized with plausible deniability to accomplish fait accompli in all these cases.[39] Post the Crimea referendum this fait accompli was legitimized.[40] Though Russia maintained plausible deniability, deniability was not essential to such a conventional strategy and depends more on the context or what the aggressor prefers. In fact, the use of hybrid warfare wasn't as consequential as the use of high-end conventional forces and capabilities like large-calibre artillery, MLRS systems, and tanks was in Crimea.[41]

The Chinese grey zone warfare works in a similar manner, but contrary to the Russian strategy there is no pretence of no involvement or deniability. What therefore matters is the overall instrumentalization of the conventional strategy that presents a fait accompli such that the onus of conventional escalation rests on the adversary. In fact, in East Asia varied forms of peacetime coercion cater to multiple conflicts and actors. China's People's Armed Forces Maritime Militia (PAFMM) poses a challenge with "salami-slicing"[42] as a coercive method to change the strategic realities on the ground rather than at sea.[43] China's other maritime forces include the PLA Navy (PLAN) and China Coast Guard (CCG).[44] In East Asia, US-China are involved in an extended deterrence dyad which makes use of such coercion more profitable and probable, as it increases the threshold of US intervention in any crisis situation and tests the intra-alliance coordination. Andrew Erickson notes that "PAFMM recruits ex-PLAN personnel with no fishing responsibilities to conduct operational deployments with larger vessels with mast-mounted water cannons for spraying, reinforced hulls for ramming".[45] There is no deniability to such actions and force is used without provoking a large-scale conflict. Another set of actions includes the base construction and island-building activities in the South China Sea (SCS) and East China Sea (ECS) region over contested islands like Spratly, Paracel, Scarborough Shoal, and Senkaku and declaring air defence identification zones (ADIZs).[46] These bases in turn provide for anti-area/anti-denial (A2/AD) capabilities in support of the conventional forces; the primary objective is not just to provide for conventional war capabilities but to coerce states into abandoning their rights.[47] These bases then allow for numerous PAFMM and CCG boats to operate for months[48] to prevent oil and gas activities by other states even as China itself continues with such activity in contested waters.[49] Meanwhile, the threat of Chinese conventional naval and air power provides a safe bubble for these forces to operate unhindered.[50] Sea control is therefore achieved as a fait accompli by ICS that piggybacks on a conventional deterrent and force posture with an offensive intent.[51] One former US Navy commander concedes that "The US has lost advantage throughout the spectrum of operations, from low-level interaction against China's maritime militia to higher-end conflict scenarios".[52] Importantly the conventional force posture is strengthened on account of the gains made through salami slicing such that in a conflict situation China would in fact "strike first" given its better domain awareness and closer bases from which fighter jets can rapidly deploy.[53] The gains made in ICS are always mutually reinforcing such that the end conventional posture is stronger than before capable of pursuing further gains.

In South Asia, the India-Pakistan dyad experiences ICS in a bidirectional manner. Pakistan uses militant outfits to stage attacks on military installations and civilians in the disputed territory of the state of Jammu and Kashmir (J&K). It provides ideological, diplomatic, and material support to these militant outfits. Militant actors with backing from state actors have over time inflicted major attacks on the Indian mainland that have engendered military crises. However, the conventional military posture that Pakistan has adopted allows it to actively engage Indian forces to allow infiltration of irregulars and sometimes combined teams of regular and irregulars on the Indian side, who then raise the cost of holding territory for the Indian side both in men and in material.

There is nothing sub-conventional in character about this coercion, and the conventional force posture is in fact pivotal to it. It communicates "militancy" as on ongoing behavioural fait accompli that India must consent to. Since 2008 cross-border skirmishes along the LoC have continued with varying intensity throughout the post-2008 period, and some particularly intense long bouts were seen in 2013–2014 and 2014–2015. This crossfire transpires between armed personnel but occasionally also targets civilian populations along the LoC, while the use of field guns, mortars, and heavy artillery is not ruled out. The objective of this brinkmanship is behavioural since there are no territorial gains. Pakistan presents the fait accompli in this dyad as normalization of attacks by violent militant actors and adopts a conventional posture along the LoC that furthers this form of coercion. Simultaneously violent activity by the state-backed, but deniable, militant actor provides it with justifiable diplomatic, moral, and strategic fig leaf to continue with such conventional posturing.

This dyad started with coercion as a territorial intent over time; however, the coercion from both sides has transformed into coercion linked with behavioural objectives. The earliest such attempts were made immediately post-nuclearization in 1999 when a "fait accompli"-style conventional land grab operation was conducted in Kargil peaks in the state of J&K. India responded to this by escalating the conflict vertically using conventional air power to take back lost territory. The mutual reinforcing element was absent in this operation because Pakistan's conventional posture, specifically the dearth of air support for the ground force elements at these new locations, was absent.

Afterward, attempts at conventional land-grab operations were abandoned in favour of the use of militant actors[54] that would target civilians and major urban centres, including an attack in 2001–2002 that targeted the Indian parliament in what would later be referred to as 2001–2002 crisis. Post the crisis India mobilized its troops; however, this mobilization took a long time and allowed Pakistan to counter mobilize. A ten-month-long military standoff between India and Pakistan ensued, but the assessments within India on the effective utilization of military instruments were mixed. India assessed that the long mobilization time sniffed out the offensive edge of the conventional operation. During this time, the Indian Army started working on a new operational concept referred to as "Cold Start Doctrine" grounded in a fait accompli strategy that was premised on swift mobilization and limited ingress to bargain vis-à-vis Pakistan using lost territory. However, the 2001–2002 crisis was followed by the Mumbai attacks of 2008. India did not mobilize troops but coercive diplomacy and the international pressure on Pakistan ensured that from then on, such attacks were limited to the states of J&K and Punjab and only targeted military installations. Despite this calibration the status quo was far from satisfying for India, as it incurred expenditure in both men and material.

Over time India started militarily reacting to a quantitative threshold of fatalities,[55] prompting it to retaliate after the Uri attack in 2016 and the Pulwama attack in 2019. The 2019 Pulwama attack was followed by Balakot air strikes by India to impose punishment or cost on Pakistan; however, the next day Pakistan responded with an aerial strike of its own leading to a skirmish. Pakistan's matching conventional response was in accordance with its doctrine of "Full Spectrum Deterrence" (FSD). More importantly post the skirmish a former Director General of Pakistan's Strategic Policy Division (SPD) announced that "Quid-Pro-Quo Plus" (QPQP) was a declared policy that essentially promises to go a rung up in its response to any limited conventional attack by India. To counter India's "Cold Start" Pakistan has implemented a "New Concept of War Fighting" (NCWF) strategy that seeks to locate units even closer to the border and with better mobilization to counter the speed imbibed in the Indian cold start concept.

A closer look at Pakistan's doctrine, operational concepts, namely, FSD, NCWF, and QPQP, evinces a continuous link between its conventional strategy and posture that are optimized for peacetime coercion and threaten escalatory responses. Pakistan therefore intends to use its conventional forces for punishment, and for threat of nuclear escalation[56] to instrumentalize this behavioural fait accompli. India, on the other hand, maintains a strong "offensive-defence" posture along the LoC to deter a spectrum of threats from any Kargil-type operations to low-level intrusions. In the meantime India has continued to alter the legal status quo of the J&K territory[57] under its control while threatening a conventional "cold start" operation if a major attack were to transpire on Indian territory. India's ICS instruments involve the use of diplomatic, economic, or institutional instruments, but the threat of conventional forces is still at play in its ICS strategy. While India's objectives are status quoist in comparison to Pakistan's revisionist in terms of territory, they are revisionist in terms of behavioural objectives.[58]

China and India exhibit a relatively more stable dyad in comparison to the other dyads mentioned here. However, competition, conflict, and coercion are not absent in this dyad. China and India share a border dispute along the Line of Actual Control (LAC) at twenty points where their claims overlap.[59] Since 1985, China has been demanding Indian concessions in the east, specifically the cession of the Tawang tract. Since China now controls Aksai Chin; it denies all disputes there, claiming that the problem relates to only approximately 2,000 km, not 4,000 km as per India's claims. Though mechanisms exist for dispute resolution and there has been technical progress, the political intent to actually settle the dispute has been lacking, thus showing that revisionism[60] is the political objective.[61] Over time China has used salami slicing and lawfare tactics to effect coercion more specific to the theatre of confrontation. Border issues have led to some transgressions as perceived by both sides, and some serious face-offs have ensued, which include Depsang in 2013, Chumur in 2014, and Doklam in 2017.[62] The bucket of coercion here involves aggressive patrolling combined with military facility construction of semi-permanent or permanent nature. On certain occasions this also included the destruction of minor military posts.[63] Border incursions are often stage-wise development which starts with sending grazing parties into claimed territory followed by overt military activity and construction.[64] Similar to China's strategy in East Asia the objective of this coercion is mutually reinforcing with the conventional military posture. The military posture both enables such tactics and is in turn reinforced by the newly acquired advantage on the ground. Doklam was a perfect iteration of this tactic where India and China had a 72-day military standoff.[65] It was prompted by China's road construction activity of permanent nature[66] at the tri-junction area with Bhutan. Assessing the significance of the area, on 18 June 2017, Indian troops moved in to form a human wall to stop China from building the road that would eventually enable it to move vehicles to south Doklam.[67] The standoff ended when both sides agreed to pull back forces,[68] with the understanding that status quo had been achieved.[69] However, satellite imagery analysis later suggested that China continued with construction activity,[70] including roads and a four-bay garage to beef up its military capacity in the immediate vicinity.[71,72] All through the Doklam standoff, the prospect of escalation was threatened by the Chinese.[73] Incidents of stone pelting were also repeated in 2018[74] and 2019.[75] Though below the threshold of conventional military action, this form of violence involving direct state military forces highlights that violence and threat of violence are continuously used to effect peacetime coercion. In fact, the Chinese have been building up their conventional capability in the vicinity to have a mutually reinforcing advantage if Chinese troop access the ridge overlooking the Siliguri corridor.[76] Amidst this ongoing faceoff another standoff took a violent turn when in June 2020 Indian and Chinese soldiers clashed at Pangong lake, Ladakh.[77] This crisis is ongoing, and although there was a limited withdrawal by both sides from the frontlines, over the next two years, the Chinese reinforced their conventional posture by using these territorial gains to build forces in the operational deep area, strengthening frontline positions by occupying critical patrol points, and setting up new bases.[78] Small outposts over time evolved into permanent bases and logistics that can quickly replenish supplies both in men and in material.[79] These can support existing troops as well as allow for surge capacity for a quick offensive. India's reaction to such coercive tactics has been to gradually move from a defensive posture based on denial to one of "deterrence by punishment".[80] This shift is apparent in its raising of new units for counter-offensive action.[81]

Role and nature of instrumentalized conventional cross-level strategy

Table 1.2 summarizes trends of coercion and "conventional deterrence" stances. Despite deterrence being the framework, the instrumentalized role of the conventional strategy is clearly visible. Compellence through ICS conveys a cost-benefit form of deterrence at each step post the completion of the forceful action, by threatening greater cost and promising as well as ensuring large-scale engagement of conventional military forces. To go back to Figure 1.2's bracket of mutual hurt between A1 and A2, the common space in which mutual hurt would operate for both adversaries is largely dependent on the credibility of threat from each other and where they don't want military action to take place. In a nibbling strategy that effectively presents small territorial gains as fait accompli,[82] the credibility of higher-scale punitive threats from the adversary is reduced,[83] such that the common hurt operates more in the higher-order conventional response spectrum (Figure 1.4). This problem is acute in extended deterrence dyads, where the challenger's interests are in fact more vital than that of the extended deterrer's commitment to the alliance.[84] This relates to a problem of "credibility of intentions" rather than "credibility of capability".[85]

The consequence of such peacetime coercion is evinced in the proliferation of punitive conventional strategies because denial strategies are not sufficiently (at least from the deterrer's perspective) successful in defeating ICS. These punitive strategies impact the bracket of mutual hurt (Figure 1.4), because they reduce the hurt bracket A2min-A2max that is now more favourably disposed to risky higher-order conventional responses, and to respond to this NWS actor A1 adjusts its A1min-A1max by decreasing its area of hurt over even higher-order conventional responses and the nuclear realm. The overall consequence is a reduction of A2min-A1max which is essentially responsible for bilateral deterrence stability. In practical terms, a reduction in the deterrence at both the lower and higher levels of conflict spectrum is witnessed, and the scale of diminished deterrence is directly proportional to the scale of punitive strategies chosen both in furtherance and in resistance of ICS.[86]

The ICS objective and the conventional posture are almost always mutually reinforcing, strongly exhibiting a linkage between the two. This linkage is established by strengthening the conventional military posture in support of the ICS objective such that the ICS gains can't be reversed without engaging a much broader set of conventional forces and perhaps over a prolonged duration. The newer assets, whether territorial or behavioural acquired, are likely to be acquired precisely because they are tenable and can further fortify the existing conventional military posture. Whether it's a case of a direct deterrence dyad or an extended deterrence dyad where the stronger partner in the alliance is committed to responding, once a fait accompli is accomplished, the countering NWS has the onus of escalation to change the established status quo. Since the new prize acquired by the aggressor is mutually reinforcing with its conventional posture, this introduces a costly "onus of escalation" factor into the defender's calculus. This calls into question the credibility of the defender NWS's conventional capability and his determination to undertake a costly, mostly long and mutually damaging, attritional exercise with the adversary, testing his will, capacity, and capability.[87] Fulfilling the deterrent threat requires bearing a much more prohibitive cost like leading to nuclear escalation as compared to the immediate loss by accepting the fait accompli.[88]

Some have referred to the nature of this problem as "coercive nuclear escalation"[89,90] however, the concept described here and the evidence from the dyadic interactions show that nuclear coercion is not instrumentalized the same way in localized, gradual, and sustained conflicts and particularly towards ICS objectives in the way conventional capability is instrumentalized. Crucially "nuclear coercion" remains limited to threats and statements,[91] whereas conventional military posture is directly utilized and is mutually reinforcing with the process of ICS. This argument is based on the "inherent escalatory nature" of conventional operations in modern times and therefore the inevitable risk of nuclear escalation that allows for the concept of "coercive nuclear escalation". This is based on the logic that increases in remote-sensing capabilities and sensory-data revolution allow such campaigns to be processed far easily. Other scholars have, however, argued that the fundamental conditions that allow the defender's land forces to fend off aggressing sea and air forces have in fact been strengthened because of this revolution.[92] The risks of nuclear escalation do exist, but this is mostly inadvertent in nature, rather than the advertent aspect implied by the "inherent" connotation. Since nuclear coercion can be "coercion" only with advertent actions and not inadvertent actions, and there are quite a few conventional "firebreaks" before advertent actions and nuclear coercion could come into play, that shows the conventional nature of such coercion.[93] In terms of advertent escalation, though nuclear deterrence is certainly involved in the sense it is essentially responsible for the limited and gradual nature objectives in the conflict, the jump from conventional to strategic nuclear realm is not an easy or unhindered transition.

This is because conventional military options are instrumentalized for escalation control or intra-war deterrence, before any form of nuclear coercion can kick in. For instance, in 2015 and 2016 Russia's deployment of S-400 and S-300 air defence systems in Syria forced the United States to abandon air operations and cruise missile strikes, allowing the Assad regime to survive, and changed the course of the conflict.[94] Here escalation control was accomplished by changing the cost-benefit calculus of the adversary before raising the prospect of terror (nuclear coercion). Such a strategy is in line with cross-level instrumentalization of conventional capability and crucially exhibits a thorough understanding of adversary thresholds.[95] ICS that inherently relies on the mutually reinforcing

TABLE 1.3 Summary of nature/scale attributes of crises in the third nuclear age.

Military Crisis Involving NWSs	*Scale Small-Unit Action/Local/ Regional/Strategic and Number of Engagements*	*Single Theatre/ Cross-theater*	*Target or Engagement Type*	*Duration Prolonged/Short*[a]	*Vertical Escalation/ Horizontal Escalation*	*Conventional Spectrum Bound High/Low*
India-Pakistan (Uri attack)[b c]	**Small Unit, 2**	**Single Theatre**	**Proxy Actor**	**Short (18, 28 September 2016)**	**Vertical**	**Low**
India-Pakistan (Pulwama-Balakot)[d]	**Limited, 3**	**Single Theatre**	**Proxy Actor/ Military Actor**	**Short (14, 26, 27 February 2019)**	**Vertical**	**High**
India-China (Doklam)[e f]	**Local, –**	**Single Theatre**	**Military Actor**	**Prolonged (73 days)**	**Vertical**	**Low**
Russia-NATO/US (Crimea)[g h]	**Limited, –**	**Single Theatre**	**Military actor with plausible deniability**	**Prolonged (27–28 February 2014 to 4–24 September 2014)**	**Horizontal**	**Low**
US-Iran[i] (Drone-strike-Missile Strike 2020) [j]	**Regional, 4**	**–**	**Military Actor**	**Short (27 December 2019, 31 December 2019, 3 January 2020, 8 January 2020)**	**Vertical**	**High**

[a] Prolonged and short are distinguished by crisis that continue for over 20 days or less than 20 days respectively. In addition, short crisis are marked by iterative engagements rather than continuous military engagements.

[b] Joy Mitra, "Post-Uri Strikes: Deterrence, Threshold, and Escalation", South Asian Voices, October 6, 2016, https://southasianvoices.org/post-uri-strikes-deterrence-threshold-and-escalation/

[c] "Transcript of Joint Briefing by MEA and MoD (September 29, 2016)", Ministry of External Affairs, India, September 29, 2016, https://www.mea.gov.in/media-briefings.htm?dtl/27446/Transcript_of_Joint_Briefing_by_MEA_and_MoD_September_29_2016

[d] Uri and the Balakot strikes, as well as Pakistan's counter-escalation all ingresses were made across the LoC rather than the International Border suggests that a horizontal threshold was in play. See Joy Mitra, "Taking Stock a Year after Balakot: India's Crisis Response and Takeaways", South Asian Voices, March 10, 2020, https://southasianvoices.org/taking-stock-a-year-after-balakot-indias-crisis-response-and-takeaways/

[e] Oriana Skylar Mastro & Arzan Tarpore, "Countering Chinese Coercion: The Case of Doklam", War on the Rocks, August 29, 2017, https://warontherocks.com/2017/08/countering-chinese-coercion-the-case-of-doklam/

[f] "Sino-India Doklam Standoff Ends after 73 Days, Ahead of BRICS Summit in China", Tibetan Review, August 29, 2017, https://www.tibetanreview.net/sino-india-doklam-standoff-ends-after-73-days-ahead-of-brics-summit-in-china/

[g] Michael Kofman, "Russian Hybrid Warfare and Other Dark Arts", War on the Rocks, March 11, 2016, https://warontherocks.com/2016/03/russian-hybrid-warfare-and-other-dark-arts/

[h] "Ukraine Crisis: Timeline", BBC News, November 13, 2014, https://www.bbc.com/news/world-middle-east-26248275

[i] "Uncomfortable Lessons: Reassessing Iran's Missile Attack", Center for Strategic and International Studies, 2020, https://www.csis.org/analysis/uncomfortable-lessons-reassessing-irans-missile-attack

[j] "US-Iran Tensions: Timeline of Events Leading to Soleimani Killing", Al Jazeera, January 9, 2020, https://www.aljazeera.com/news/2020/01/iran-tensions-timeline-events-leading-soleimani-killing-200103152234464.html

nature of the conventional posture and newly gained assets has escalation control or intra-war deterrent ingrained in its operationalization from the very start, and once the gains are made, it pivots to escalation control quite rapidly.

A second way in which conventional escalation control is operationalized is by limitation which is evinced across crises in this period (see Table 1.3). Conventional escalation control is operationalized, by the absence of one or more of the following factors: avoidance of (1) large-scale engagement, or (2) direct engagement of military targets; (3) protracted conflict engagement; (4) horizontal escalation; (5) higher-order conventional responses. These thresholds are easily understood by both adversaries and are visible, and all of them are crucial to allow space for de-escalation. Since the conventional actions that occur between NWSs are mostly at the periphery, deliberate escalation is effected without conveying an unlimited political or military intent and without provoking a response that is unlimited in nature. The limitation in intent in one or more of these factors is a mechanism of assurance from the conventional deterrer/compeller to the adversary. In some cases, even when high-order conventional responses or direct engagements of military targets are undertaken against adversaries, these engagements are limited to a single iterations and protracted engagement is avoided. Alternatively, this implies that conventional deterrence/compellence objectives can therefore be achieved/established gradually over iterative engagements rather than in one fell swoop.

Here it is also important to note that if, say, the character of the conflict spectrum changes, due to an expansion of the CN min-CN max space, that is, the conventional-nuclear overlap expands due to technological changes, then the graduation of conventional strategy from the role of a peacetime coercive instrument to that of an escalation control or intra-war deterrent will also kick in earlier in the military crisis. This means that the progression of a crisis from something at local or tactical level to that involving operational-level forces and ramifications is also going to be swift. The risks of escalation are therefore palpable even in a small crisis as conventional strategy rapidly pivots from peacetime coercion to escalation control/intra-war deterrence. This is discussed in greater detail in later chapters.

A direct ramification of ICS is that at least at the initial stages of the conflict the prospect of vertical escalation becomes more probable than horizontal escalation, as the conventional strategy in support or resistance of ICS objective seeks to intensify the use of violence localized to the particular theatre of conflict to re-establish the perceived gap in deterrence. This is also because deterrence is essentially a function of time and location. When we say that a state actor, say, India, deters China, it does not imply that it deters it across all geographical locations. India may be able to deter China in the Indian Ocean but might not be able to always deter it along the land borders. Similarly, the United States may have effective deterrence against China in the South China Sea but may not be able to do that all throughout the year if there are multiple crises that demand limited military resources be allocated elsewhere. Essentially national-level notions of conventional deterrence must give way to conventional deterrence capacity in a certain geographical theatre. The fact that deterrence is essentially local and time-specific in nature is also a reason why it is also much more likely to break down between nuclear-armed states in the TNA. The assumption that conventional conflict cannot take place between nuclear-armed states as it will lead to mutually assured destruction (MAD) is challenged in this new paradigm because conflict can be iterative and limited in time and space at specific locations, where effective deterrence is not a constant, but a dynamic, interplay of various factors.

Lowered nuclear threshold: first use or re-use?

The conventional realm is bounded at (CN max) and is set by the principles of avoiding nuclear annihilation and therefore avoiding nuclear warfighting, but whether this bound is crossed or not is set in the political domain. The lower bound of the purely nuclear spectrum, the point at which a conflict "transitions" from conventional to nuclear, sets the distinction between the realms of conventional and nuclear strategy. This point is concerned with advertent or deliberate escalation that precipitates in the use of a tactical (or battlefield or non-strategic) nuclear weapon.[96] Deliberate use of a TNW and its aftermath situation is critical in determining to what extent the conventional spectrum and thereby conventional military strategy have a role to play.[97] That vertical escalation is more probable makes it likely that a demonstrative or militarily beneficial[98] use of TNWs in a conflict that is localized could take place.[99] TNW usage is after all the single most important act of intra-war deterrence in a conflict, and one where the state using TNW signals the prospect of "terror" rather than "cost-benefit" calculus.

The US DoD defines the use of tactical (or non-strategic or battlefield) nuclear weapons as

> "the use of nuclear weapons by land, sea, or air forces against opposing forces, supporting installations or facilities, in support of operations that contribute to the accomplishment of a military mission of limited scope, or in support of the military commander's scheme of manoeuvre, usually limited to the area of military operations."[100]

Strategic and non-strategic nuclear weapons were also distinguished on the basis of range of the delivery systems to hit within a nation's territorial bounds, but this distinction is a tough one to make in the case of contiguous nuclear weapons states or dual-capable aircraft.[101] Another criterion could be to distinguish them on the basis of the yield; however, modern systems again blur this distinction on account of their accuracy which, if intended, could achieve "strategic effect" in a conflict.[102] Still others consider all nuclear weapons strategic regardless of the delivery or yield attributes. For instance, India, at least at the declaratory level, makes no distinction between strategic and non-strategic nuclear use. Therefore, the dilemma surrounding what is tactical or strategic and conventional or nuclear is often less about terminology and more a function of the targeting strategy.[103]

In classical literature TNW use[104] isn't merely tactical because in essence they have the strategic effect of transitioning the conventional conflict over to the nuclear realm.[105] However analysts have pointed out that the nuclear taboo, as it is referred to, is less a result of shared "norms" and more of a deliberate, product of practical calculations about battlefield utility of TNWs[106] and the domestic calculus about national interests of the state.[107] This calculus around usage of TNWs has particularly shifted in the TNA because ICS has led to a proliferation of conventional strategies and postures that serve compellent or punishment ends; in particular from the adversary's perception most actions are an exercise in compellence. They simultaneously pressure both "first use" nuclear doctrines or asymmetric escalation nuclear postures, and second, those based on "massive retaliation" or assured retaliation posture to push the conflict into the nuclear realm.

The introduction of TNWs in the arsenal is based on the assumption that the concept of "limited nuclear war" exists.[108] They restore some of the lost conventional parity in a way that introduces the "balance of terror"[109] aspect at the conventional level[110] without bringing in the question of nuclear superiority. Particularly radiation weapons or neutron bombs in fact combine a unique set of characteristics of low-yield, small blast radius, and heat effects that, if detonated sufficiently above ground, make them conducive for battlefield usage.[111] Historically, during the Cold War, the United States was inclined to use TNWs in conjunction with conventional weapons to target important command and control, logistic facilities, and troop concentrations,[112] while the Soviet strategy included theatre nuclear strikes in conjunction with large-scale ground operations.[113] The existence of "use or lose" pressure, however, contradicted the very concept of limited tactical use.[114] In both cases forward-deployed short-range TNWs had to be necessarily assigned to a lower level of command, making them vulnerable to adversary strikes while decreasing control over their usage.[115]

These factors call into question the military utility of the weapon. This, however, did not rule out the demonstrative value of the single shot[116] or its role as an intra-war deterrent. Advertent singular detonation could clearly communicate the next stage of escalation. However, it is entirely possible that mandates of escalation dominance and damage limitation could force the adversary to raise stakes and skip rungs respectively, or that such use could be construed as weakness either by the initiator or by the receiver.[117] The increased prospect of vertical escalation implies that actors could consider the feasibility of such a limited nuclear exchange. Klaus Knorr, for instance, says that limited nuclear exchange would make sense, "if limited war capabilities are insufficient to contain local aggression" and that the feasibility and utility of this limited nuclear exchange would depend on the "degree to which strategic forces were vulnerable to counter-force attacks".[118] This is consistent with the logic of using unlimited military means for a limited political aim strategy rather than the use of limited military means for unlimited political aims.[119]

Tactical usage of nuclear weapons may therefore result from either failure of the conventional campaign or because of the desire for freezing the current politico-military ground situation.[120] At the outset the conflict between NWSs is almost always limited in its political objectives as no one is looking for a costly decisive victory that is certain to enter the nuclear realm. However during the course of the conflict, if the outcome is an attritional or an unfavourable stalemate, then the defender/aggressor may contemplate TNW usage to freeze the status quo before running out of resources in men and material. Conventional tactics can certainly be modified to incorporate nuclear contingencies and operations in post-nuclear detonation environments to absorb the battlefield impact of TNW weapons.[121] This gives way to probable scenarios where the recipient after "first use" by the initiator could consider conventional-only or nuclear strategy responses.

Vincent and Aaron write that "a strategy of continuing the conventional campaign toward victory after adversary limited nuclear use would likely provide the enemy with ample time and incentive to employ additional nuclear attacks".[122] To that end they recommended underpinning a conventional-only response with credible nuclear options.[123] The other factor to contend is that if the conflict is to be limited or localized, continuing conventional operations without taking out the adversary's command and control infrastructure[124] or suppressing his air defences may lead to significant infliction of damage or further nuclear escalation from the adversary. Whether the conflict enters the purely nuclear realm or not at this stage is therefore an outcome of mutual capabilities and decision-making that is concerned not just with first use but re-use.

As a consequence of endeavours to plan for conventional-nuclear transition,[125] TNWs are a necessary part of conventional strategy and operations. With the idea of escalation management and "restoring deterrence following nuclear first use" conventional military strategies must distinguish between deterring first use[126] and deterring "subsequent use of nuclear weapons".[127] In that sense the conventional military strategy paradoxically adds a conventional dimension to "limited nuclear war" concept. This is based on the logic that even after a TNW is used, it does not necessarily cross the framework of "coercion" to enter into "brute force"[128] as persuasion and intimidation are still involved in its use and with the potential threat of re-use. The first scenario could end in re-use of nuclear weapons; this could happen if the recipient state responds in kind with nuclear weapons to push the conflict in the realm of nuclear strategy but limited to the theatre. The second scenario could have the recipient not respond with nuclear response but continue with a conventional-only response, where potential for coercion, brute force, and re-use remains. Third, if conventional operations continue and this materializes into TNW re-use, then the recipient must contend with greater pressure to nuclear counter-escalate or continue conventional operations while taking nuclear hits. Finally, coercion would have been realized if conventional operations halt with the demonstrative or first use of TNW. This has two results: one planning of conventional operations must take into account the contingencies that may drive the adversary towards TNW use;[129] second, provisioning of conventional options after use of TNW introduces essentially a conventional dimension to the "'limited nuclear exchange". More importantly, nuclear strategy may not necessarily and exclusively take over conventional operations at "first use".

However, for both the initiator and the receiver the attempts at designing a force posture and doctrine that can effectively operate in nuclear or non-nuclear environments could be difficult as in the past this has tended to end in a unit that can do neither of the two.[130] This choice is tyrannical in terms of force design, as one could either employ dispersion in the battlefield to survive in a nuclear contingency or concentrate for mass to actualize a blitzkrieg operation on a modern conventional battlefield.[131] There is no balance in between. This has implications for both the aggressor and the defender. Typically, an aggressor in a conventional attrition strategy needs a force ratio of 3:1, while in a blitzkrieg strategy more prevalent on the modern battlefield where the axes of advance are limited to specific vulnerable points, this force ratio goes up to 5:1.[132] In a battlefield the aggressor must take into account that massed forces in a blitzkrieg-style penetration will present an attractive concentrated target for TNW use. This problem is also exacerbated by the fact that for the localized theatre-specific objective at least in the initial phase of the conflict defence is likely to be saturated,[133] thus reducing the prospect of a breakthrough.[134] Second, if the defender's rear isn't very deep it will trigger TNW usage fairly early in the conflict.[135] However, if the defender intends to offer conventional resistance initially, he must contend with the same dilemma; if his conventional posture depends on heavy concentration and saturation on a limited front, then post 'first use' his forces will also present an attractive target for the adversary. Another problem is that such use could hurt his own forces who are likely to be in close proximity to the aggressors.

In the India-Pakistan dyad, for example, recent Indian military exercises[136] suggest a posture of continuing conventional operations even in a post-detonation environment or taking out TNW batteries before they could be used.[137] This is coupled with improved mobile missile defence capabilities,[138] particularly potent at the tactical level to ward off the threat of TNWs, suggesting that conventional operations will be the preferred options. Pakistan may then have to contend with the possibility of TNW re-use which might trigger Indian nuclear retaliation. However, the tyranny of geography in this dyad obfuscates the strategic and non-strategic distinction in more than one way.[139] Pakistan's lack of geographical depth and the alignment of all its major industrial cities and population centres along a longitudinal axis parallel to the border with India make conventional operations threatening to the very survival of the state.[140] Second, even short-ranged delivery systems could easily be considered to deliver "strategic effect" given the kind of targets that may be within reach. In table-top exercises Pakistan scholars have not had a single opinion about whether taking out

Pakistan TNWs by conventional means would constitute nuclear escalation.[141] For these reasons the threshold for first use in this dyad is very low. Despite its official stance, India is not bereft of its own tactical nuclear options.[142] A flexible response strategy, however, seems to be the way as the Indian nuclear doctrine considers nuclear retaliation for attack on "Indian forces anywhere",[143] while its military exercises and official statements[144] suggest continued conventional warfighting options.

In the US-Russia dyad analysts[145] are divided over the Russian threshold for nuclear use. US NPR 2018 posits a lowered threshold for Russian nuclear use; however, Russia's own military doctrine states that the condition for such usage is an "existential threat". In recent years Russian conventional and non-nuclear capabilities along with superior operational-level planning[146] have also closed the gap between itself and the United States in the conventional realm. US perception of lowered Russian nuclear threshold has pushed it to field a low-yield Submarine Launched Ballistic Missile (SLBM) warhead and Submarine Launched Cruise Missile (SLCM) to provide "deterrent effect" and "without relying on the host nation". US NPR states that this closes the "mistaken perception of an exploitable 'gap' in US regional deterrence".[147] Putting low-yield nuclear warheads upon long-range delivery systems suggests a posture of nuke for nuke, rather than continuing conventional operations underpinned by a nuclear threat to deter nuclear re-use. More importantly putting a tactical warhead upon an SLBM blurs the distinguishability between strategic and non-strategic weapons.[148] This diminishes the space for effecting the threat for re-use, as first use itself could evoke a strategic response, whether damage limiting or unlimited, thereby ending the space for conventional strategy and military means. At the same time though Russia fields a strong non-strategic nuclear force[149] along with attendant delivery systems, the "existential threat" contingency is unlikely to arrive. During the Cold War period, the Soviet Union's TNWs had a much greater role in operational plans in the theatre of conflict to reinforce conventional units in large-scale land or sea operations.[150] Of late some Russian military exercises have simulated the use of nuclear weapons against NATO members. But the latest doctrine relegates the role for nuclear weapons. In this dyad the threshold for nuclear use is high, and the potential for strategic nuclear warfighting is also high.

In the US-China dyad, Chinese maintain an unambiguous no-first-use policy.[151] However, the United States is concerned about its theatre nuclear-capable missiles, as the Chinese maintain superiority in the number of such delivery systems and such a stockpile can technically be used against US bases.[152] This perception is strengthened on account of a classified Chinese training manual[153] that allegedly suggests "lowering of nuclear threshold" when faced with extreme conditions; however, the lowering here implies a higher alert level for nuclear weapons[154] rather than actual first use of nuclear weapons. The Chinese are, however, more explicit about how they see nothing "conventional" about precision-guided conventional systems for conventional counterforce, which in fact in their view would constitute nuclear first use.[155] However analysts have argued that Chinese theatre nuclear capability is relatively invulnerable to US conventional counterforce.[156] Alternatively, the United States could consider first use of nuclear weapons if its conventional capability erodes[157] significantly in the opening round of a regional conflict.[158]

Finally, in the India-China dyad the military intervention threshold appears to be fairly high. Though cross-level conventional coercion takes place, it is with the threat and risk of conventional conflict that is localized or limited. Chinese threat perception regarding India is also low in comparison to other adversaries, although there is some concern regarding India's conditional NFU and if there is a greater role assigned to the nuclear force than just strategic deterrence.[159] As both NWSs pledge NFU at the declaratory level, the use of a demonstrative tactical nuclear weapon as an advertent escalatory response is ruled out. However, the Chinese conventional strategy based on "strike first" logic and its heavy dependence on conventional ballistic missiles could possibly compel India to review its NFU commitment especially if its conventional capability is eroded significantly in the opening round of the conflict, to serve as an instrument for intra-war deterrence.[160] First use is likely to occur if conventional operations don't cease even after review of the NFU commitment, post which theatre nuclear warfighting could take place.

Overall, the reason TNW use gets introduced in the calculation is because fundamental to ICS is protecting gains with a larger mutually reinforcing military posture. Nuclear posture is designed to create options post the intra-war deterrent stage and caters to this space post the exhaustion of conventional escalation control options. This effectively means that apart from the conventional posture playing the role of the intra-war deterrent, TNWs also cater to an intra-war deterrent role, albeit at higher level in the conflict and effectively share that space with conventional military forces and posture. This is distinct from the role of the nuclear posture which is utilized in deterring existential threats to the state. This invites TNW use if the conflict expands in scale, vertically or horizontally beyond the conventional escalation control options. Evidencing is the fact that in most cases TNWs exist in the force structure without a doctrine that really specifies the conditions for their early use on the conventional battlefield which makes it difficult to predict the exact event that could precipitate in their use. The prospect of advertent vertical escalation therefore suggests their demonstrative utility in communicating potential strategic nuclear escalation. This transition point from conventional to nuclear courtesy advertent escalation is more fluid in some dyads than others. In most cases the receiver either has a flexible response strategy or is moving towards one to respond to a TNW "first use". Conventional strategies and operations therefore necessarily incorporate plans for nuclear contingencies and environments as both recipients and initiators. This takes us to the heart of TNA where the propensity to use conventional forces for compellent objectives comes along with the attendant risk of nuclear use.

ICS template

ICS instrumentalization of conventional military towards the objectives of peacetime coercion, escalation control or intra-war deterrence, and finally integrating tactical nuclear weapon in the conventional military strategy does beg a discussion on how this template might interact with different doctrinal attributes and postures. In a generalized case, say, NWS A1 initiating compellent action through ICS, it puts the onus of responding/escalation on NWS A2. As a result of A1's actions A2's conventional posture has the onus of escalation, courtesy the very nature of ICS that is inherently based on a mutually reinforcing conventional deterrent posture and conflict limitation in scale, iteration, duration, or engagement type. A1's conventional intra-war deterrent communicates a cost-benefit form of deterrence and conflict limitations that smartly test A2's thresholds for military action while a strong A1 conventional posture discourages a strong response from A2.

The choice for A2 was to maintain a deterrence by denial or deterrence by punishment posture, both of which have input costs. Military infrastructure, terrain type, and the cost of maintaining denial versus punishment posture are all variables that decide the eventual posture. Some geographic theatres with mountainous or hilly terrain might be more suitable for denial strategy, while others like maritime zones or islands might be more suited for an offensive or punishment strategy. A denial strategy has the advantage that it reverses the "onus of escalation" on A1 instead of A2; this interdicts the ICS cycle at the beginning itself. However, this strategy is often prohibitively costly and invariably in certain kinds of terrains logistical challenges may make such a posture impossible to sustain. However, if the terrain favours defence, then denial can be a better bet given that it takes away the advantage of surprise or onus of escalation, fundamental to ICS. However, a denial-based posture also means that one may have to spread their forces thin in order to achieve a pure denial posture, often neither possible nor advisable.

The other option is punishment which could take the form of dislodging at the site, intra-theatre coercion or inter-theatre coercion. Of these dislodging at the site, though likely to avoid the nuclear intra-war deterrent, is certainly a more costly option. Dislodging, though limited by nature, has to contend with the fact that the acquired asset is mutually reinforcing with the existing conventional posture of the ICS initiator. The other option is intra-theatre coercion, in which case A2 may conduct military action at a different site within the same theatre. Although the horizontal escalation in this case is limited, the initiator of ICS A1 after exhausting the conventional intra-war deterrent has the option to rope in its nuclear posture to cater to the role of an intra-war deterrent.

This is also applicable for inter-theatre coercion, in which case A2 can respond with military action outside the theatre of initial conflict. If A1 deduces that such action is likely, then it may be more favourably disposed towards having a nuclear posture that allows demonstrative first use. In such a scenario, especially if the conflict expands outside the theatre, conventional escalation control is likely to be passed on quickly, as a result of significant horizontal escalation. Theatre nuclear weapons might be utilized as an intra-war deterrent due to horizontal escalation. The difficulty with inter-theatre coercion is of course that it may not help in the return of asset that is lost in the theatre where ICS was initiated in the first place while also being significantly escalatory.

The 2020 India-China conflict in the Ladakh region fits this template quite well. First of relevance here is the Chinese ICS had the intent of dominating the Darbuk–Shyok-DBO Road. The Chinese used their conventional forces to intrude and capture specific locations all along the LAC in Galwan, Hot Springs, Gogra, and Pongang Tso. Despite the fact that the overall intrusion area was approximately in some 60 sqkm of territory,[161] the fact that the intrusions were at multiple points increased the scale factor of the ingress. However, this was accompanied by a strong mutually reinforcing nature of the conventional posture. These points of intrusion were chosen precisely because they carried military value.[162] This was reinforced with the creation of military infrastructure in the area, and crucially by inducting more Chinese division-level units in the area to maintain local superiority.[163]

An evaluation of the Indian conventional military posture in this episode reveals that the posture was largely a mix of denial and punishment whilst not being optimized for either of those deterrent strategies. The hilly terrain along the LAC is not conducive for large-scale offensive operations, but it rather supports static defence.[164] While the underdeveloped military infrastructure did not particularly support offence. According to some analysts the India military did not operationalize an effective denial by not protecting the ingress routes that would allow Chinese forces from occupying heights overlooking the DBSO road.[165]

On the other hand, the Indian punishment strategy was not optimized either. India had plans to develop a Mountain Strike Corps with two divisions in order to have a viable punishment strategy based on intra-theatre coercion. However, lack of funds led India to raise just one division, which too was underequipped[166] and without adequate military infrastructure that could sustain large-scale logistics required for offence. This is despite Indian military bases having mobilization advantages as compared to the PLA.[167] The PLA troop build-up both in the rear and in front areas was not responded to effectively in the beginning, resulting in a loss of local superiority while ingress routes were not protected because of a suboptimal denial strategy. This shifted the onus of escalation entirely onto the Indian side while the PLA conventional posture presented an unfavourable cost-benefit calculation for the Indian response, thus achieving a compellent "land grab". China did engage in some nuclear messaging;[168] however, this signalling did not involve any overt changes to the nuclear posture, suggesting that China was looking mainly at conventional escalation control. ICS framework would ascribe this to the fact that Chinese perceived Indian deterrent strategy as mainly denial in nature, while theatre nuclear weapon usage as an intra-war deterrent could have come into the picture if cross-theatre punishment were to be used by India. There are two important things to note here. First, the ICS initiator has the advantage of shaping the "onus of escalation" factor with the scale of military forces, the number of iterations, etc. and choosing a favourable point in time and space where effective deterrence is low. This implies that for both actors material investments into their military posture matter more than behavioural credibility. From the initiator A1's perception A2's credibility of capability and credibility of intent are not mutually exclusive, and therefore credibility of capability is definitely a more important determinant for A1. Second, A2 must optimize its deterrent strategy in favour of either denial or punishment; a mixed approach may end up doing neither of the two. While a mixed approach could be taken for intra-theatre or inter-theatre coercion, an appropriate posture suited for the primary theatre needs to be optimized to use such a strategy.

Conclusion

This chapter has treated the conflict spectrum in a linear fashion; in reality while the conflict spectrum does not have a linear character, inadvertent escalation and nuclear misperception are a constant threat that need to be balanced out versus set conventional strategy plans and objectives. This inadvertent factor is embedded in the conflict spectrum space and is a function of both strategy and technological entanglement in nuclear and conventional realms and is discussed later. However, the linear treatment does allow teasing out the aspect of how advertent escalation is packed into conventional strategy and the coercion role, scope, and objectives that are afforded to conventional strategy across NWS dyads. This is true for all "grey zone" and "hybrid warfare" strategies where conventional forces are in fact fundamental to the instrument of coercion being employed.

At one end conventional strategy is instrumentalized towards peacetime coercion, while at the other it transitions or merges into the nuclear level of the conflict spectrum. This cross-level character of conventional strategy is essentially responsible for what is being referred to as the third nuclear age, where compellence objectives instrumentalize conventional strategy. In addition, the fact that effective deterrence is not a constant, rather it is dynamic and is essentially local in the TNA allows more frequent deterrence breakdown between NWSs. This has the effect of making all deterrent strategies even those operationalized in defence appear and serve compellent objectives a key feature of the TNA.

Though the cross-level character of conventional strategy, on one hand, serves as a conflict initiator and on the other hand has the goal of intra-war deterrence or escalation control and there is some tension in these aspects, it is invariably resolved by according primacy to the compellent objective, such that if the objective or fait accompli is established, it improves the ground situation in favour of the existing conventional posture because the newly acquired gains are mutually reinforcing with it. ICS initiator can then deter and defend the newly acquired status quo and puts the costly onus of escalation on the defender. Such strategies necessarily depend on a thorough understanding of adversary thresholds, as well as exploring any kind of ambiguity, whether political, military, or theatre-specific thresholds, that can confuse the adversary decision-making cycle.

Therefore, theatre-specific capabilities become very important. While some theatres might be more suited to land-centric forces and capabilities, others may require a different approach, the force projection necessarily needs to be more local. Under ICS, horizontal escalation is unlikely to transpire initially, and a cross-theatre response is more likely to take the conflict beyond conventional escalation control. Therefore cross-theatre capability is likely to be less effective than single theatre–centric capabilities in achieving an effective deterrence, particularly as the asset of strategic value once lost in a fait accompli action may not be recovered in the primary theatre. This is especially true on land where it is easier to implement ICS but increasingly also true for the maritime theatre as shown by China in the SCS. Finally, as prospects of vertical escalation are more likely, conventional strategies and operations must therefore contend with the question of advertent demonstrative use of TNW and its aftermath where the nuclear doctrine or postures also cater to serve the intra-war deterrent space once conventional escalation control ends. The next chapter will deal with how some of the emerging concepts and technologies fit in the two roles of ICS, namely the peacetime coercive intent and that of conventional intra-war deterrent.

Notes

1 Thomas C. Schelling. "Arms and Influence". Yale University Press, 1966. www.jstor.org/stable/j.ctt5vm52s. This text uses the term "mutual hurt" to better capture the essence of how the ability to hurt to each other is responsible for deterrence.
2 Quoted in Schelling, Thomas C. "Arms and Influence". Routledge, 2008. Strategic Studies, pp. 96-114, pg. x.
3 See Glenn H. Snyder. "Deterrence and Defense". Princeton University Press, 1961. doi:10.2307/j.ctt183pj49. pg. 12.
4 Austin Long. "Deterrence -- From Cold War to Long War: Lessons from Six Decades of RAND Research". RAND Corporation, 2008. https://www.rand.org/pubs/monographs/MG636.html. pg. 7.
5 NWS here refers to nuclear armed states those recognized by the Non-Proliferation Treaty (NPT) as well as those who are outside the ambit of NPT.
6 Implying use of strategic nuclear weapons.
7 Biddle writes that the central characteristic of all forms of coercion is that they depend, ultimately, on cooperation by the party receiving the threat. This is by no means friendly cooperation, but it is cooperation nonetheless and therefore acts as a mutual leverage. See Tami Davis Biddle, "Coercion Theory: A Basic Introduction for Practitioners", Texas National Security Review, 20 February 2020, https://tnsr.org/2020/02/coercion-theory-a-basic-introduction-for-practitioners/
8 Nicholas L. Miller & Narang Vipin, "Is a New Nuclear Age Upon Us?", Foreign Affairs, 30 December 2019, https://www.foreignaffairs.com/articles/2019-12-30/new-nuclear-age-upon-us
9 Ibid.

10 Deterrence is essentially preventive in nature and takes the form of denial or punishment. Deterrence by denial closely hinges on the local military balance but is not the same. Deterrence by punishment raises the cost for the adversary by inflicting damage not merely limited to the defence of the local theatre of conflict. See Michael J. Mazarr. "Understanding Deterrence". RAND Corporation, 2018, https://www.rand.org/pubs/perspectives/PE295.html
11 According to R. Brad non-nuclear capabilities, missile defences and conventional strike substitutes for nuclear weapons are viewed as good complements but not as substitutes for nuclear weapons. See Roberts Brad, "U.S. Nuclear Weapons in the Twenty-First Century", Carnegie, 2016, https://carnegieendowment.org/files/Transcript17.pdf. pg. 4.
12 For more on this see T. Negeen Pegahi, "The New India Versus the Nuclear Revolution: The Future of Crises Among Nuclear Powers", War on the Rocks, 12 June 2019, https://warontherocks.com/2019/06/the-new-india-versus-the-nuclear-revolution-the-future-of-crises-among-nuclear-powers/
13 See Tami Davis Biddle, "Coercion Theory: A Basic Introduction for Practitioners", Texas National Security Review, 20 February 2020, https://tnsr.org/2020/02/coercion-theory-a-basic-introduction-for-practitioners/#_ftn1
14 Schelling, pg. x.
15 Ibid.
16 See Dan Altman (December 2017) By Fait Accompli, not coercion: How states wrest territory from their adversaries, International Studies Quarterly, 6:4, 881–891, https://doi.org/10.1093/isq/sqx049
17 Brad Robert, "U.S. Nuclear Weapons in the Twenty-First Century", *Carnegie Endowment*, 2016, https://carnegieendowment.org/files/Transcript17.pdf. pg. 5.
18 The text uses instrumentalized conventional strategy which is different from nuclear coercion. The differences in these concepts will be discussed later in the text.
19 Mearsheimer phrases this point cogently to say that conventional deterrence is essentially a function of the military strategy. See John J. Mearsheimer. "Conventional Deterrence". Cornell University Press, 1983. www.jstor.org/stable/10.7591/j.cttlrv61v2
20 This view is brought out by Negeen Pegahi, who argues that if the political objective needs revision of the status quo then even a hundred per cent successful deterrent strategy cannot achieve the set political objective. See T. Negeen Pegahi, "Deterrence in Retreat: How the Cold War's Core Principle Fell Out of Fashion", War on the Rocks, 7 December 2017, https://warontherocks.com/2017/12/deterrence-retreat-cold-wars-core-principle-fell-fashion/
21 Keir A. Lieber & Daryl G. Press. "Preventing Escalation during Conventional Wars". 2015. pg. 28.
22 R. Brad characterizes this concept roughly as the "availability of the military toolkit comprising kinetic and non-kinetic capabilities for all three levels of war". See Roberts, Brad. "U.S. Nuclear Weapons in the Twenty-First Century", 2016, https://carnegieendowment.org/files/Transcript17.pdf. pg. 7. Whereas Keir and Daryl refer to this as an escalatory strategy backed with nuclear coercion and call it a "strategy of coercive nuclear escalation" based on the logic that most conventional concepts depend on degradation of command and control which is inherently escalatory; see Lieber Keir A. & Daryl G. Press, "Preventing Escalation during Conventional Wars", 2015
23 Cross-domain deterrence does not have single common definition of domains. US Department of Defense (DoD) defines five domains: air, land, sea, cyber, and space. Some other scholars define it in terms of weapons and actors, like non-state actors, hybrid warfare, nuclear and conventional warfare, space warfare, and cyberwar. Cited in King Mallory. "New Challenges in Cross-Domain Deterrence". RAND

Corporation, 2018, https://www.rand.org/pubs/perspectives/PE259.html. This text uses the DoD classification.

24 Kristin Ven Bruusgaard, "The Myth of Russia's Lowered Nuclear Threshold", War on the Rocks, 22 September 2017, https://warontherocks.com/2017/09/the-myth-of-russias-lowered-nuclear-threshold/

25 Dmitry (Dima) Adamsky (2018) From Moscow with Coercion: Russian deterrence theory and strategic culture, Journal of Strategic Studies, 41:1–2, 33–60, DOI: 10.1080/01402390.2017.1347872. And Oliker Olga, "The Nuclear Posture Review and Russian 'De-Escalation:' A Dangerous Solution to a Nonexistent Problem", War on the Rocks, 20 February 2018, https://warontherocks.com/2018/02/nuclear-posture-review-russian-de-escalation-dangerous-solution-nonexistent-problem/

26 Ibid.

27 Kristin Ven Bruusgaard, "The Myth of Russia's Lowered Nuclear Threshold", War on the Rocks, 22 September 2017, https://warontherocks.com/2017/09/the-myth-of-russias-lowered-nuclear-threshold/

28 "Russian Special Dispatch No. 7943", Middle East Media Research Institute, 14 March 2019, https://www.memri.org/reports/russian-first-deputy-defense-minister-gerasimov-our-response-based-active-defense-strategy

29 Ibid.

30 John J. Mearsheimer, "Why the Ukraine Crisis Is the West's Fault", Foreign Affairs. Foreign Affairs Magazine, 18 August 2014, https://www.foreignaffairs.com/articles/russia-fsu/2014-08-18/why-ukraine-crisis-west-s-fault

31 D. Johnson, "Russia's Approach to Conflict—Implications for NATO's Deterrence and Defence", NATO Defence College Research Paper No. 111, April 2015, www.ndc.nato.int/news/current_news.php?icode=797. pg. 2.

32 Ibid, pg. 6.

33 J. Durkalec, "Nuclear-Backed "Little Green Men:" Nuclear Messaging in the Ukraine Crisis", Stratcomcoe, 2020, https://www.stratcomcoe.org/jdurkalec-nuclear-backed-little-green-men-nuclear-messaging-ukraine-crisis

34 Elbridge Colby, "For Peace with Russia, Prepare for War in Europe: NATO and Conventional Deterrence", War on the Rocks, 20 April 2016, https://warontherocks.com/2016/04/for-peace-with-russia-prepare-for-war-in-europe-nato-and-conventional-deterrence/

35 Ibid.

36 "EU Warning over Russia 'land Grab' in South Ossetia Border Row", BBC News, 16 July 2015, https://www.bbc.com/news/world-europe-33549462

37 Russia deployed some 40,000 troops on Ukraine's border in May 2014. Igor Sutyagin & Michael Clarke, "Ukraine Military Dispositions: The Military Ticks Up While the Clock Ticks Down", Royal United Service Institute, 2014, https://rusi.org/system/files/UKRANIANMILITARYDISPOSITIONS_RUSIBRIEFING.pdf

38 Michael Kofman, "Fixing NATO Deterrence in the East, or How I Learned to Stop Worrying and Love NATO's Crushing Defeat by Russia", War on the Rocks, 12 May 2016, https://warontherocks.com/2016/05/fixing-nato-deterrence-in-the-east-or-how-i-learned-to-stop-worrying-and-love-natos-crushing-defeat-by-russia/

39 Dan Altman, "The Long History of 'Green Men' Tactics—And How They Were Defeated", War on the Rocks, 17 March 2016, https://warontherocks.com/2016/03/the-long-history-of-green-men-tactics-and-how-they-were-defeated/

40 D. Johnson, "Russia's Approach to Conflict—Implications for NATO's Deterrence and Defence", NATO Defence College Research Paper No. 111, April 2015, www.ndc.nato.int/news/current_news.php?icode=797. pg. 8.

41 Michael Kofman, "Russian Hybrid Warfare and Other Dark Arts", War on the Rocks, 11 March 2016, https://warontherocks.com/2016/03/russian-hybrid-warfare-and-other-dark-arts/

42 The strategy has also been called "talk and take". See, for example, Anders Corr, "China's Take-And Talk Strategy in the South China Sea", Forbes, 29 March 2017, See also Namrata Goswami, "Can China Be Taken Seriously on its 'Word' to Negotiate Disputed Territory?" The Diplomat, 18 August 2017.
43 "A Short History of China's Fishing Militia and What It May Tell Us", 2020, http://www.maritimeissues.com/history/a-short-history-of-chinas-fishing-militia-and-what-it-may-tell-us.html
44 "The Complete China Maritime Militia Bookshelf: Latest & Greatest", Andrewerickson.com, 13 December 2018, http://www.andrewerickson.com/2018/12/the-complete-china-maritime-militia-bookshelf-latest-greatest-4/
45 Ibid.
46 Ronald O. Rourke, "Maritime Territorial and Exclusive Economic Zone (EEZ) Disputes Involving China: Issues for Congress", Congressional Research Service, 2016, https://fas.org/sgp/crs/row/R42784.pdf
47 Gregory B. Poling, "The Conventional Wisdom on China's Island Bases Is Dangerously Wrong", War on the Rocks, 10 January 2020, https://warontherocks.com/2020/01/the-conventional-wisdom-on-chinas-island-bases-is-dangerously-wrong/
48 Ibid.
49 "UPDATE: China Risks Flare-Up Over Malaysian, Vietnamese Gas Resources", Asia Maritime Transparency Initiative, 2016, https://amti.csis.org/china-risks-flare-up-over-malaysian-vietnamese-gas-resources/
50 Gregory B. Poling, "The Conventional Wisdom on China's Island Bases Is Dangerously Wrong", War on the Rocks, 10 January 2020, https://warontherocks.com/2020/01/the-conventional-wisdom-on-chinas-island-bases-is-dangerously-wrong/
51 "China's Sea Control Is a Done Deal, 'Short of War With the U.S.'", The New York Times, 20 September 2018, https://www.nytimes.com/2018/09/20/world/asia/south-china-sea-navy.html
52 "South China Sea Fury: China Now 'Outmatches US in Every Way' as Standoff Boils", Express UK, 19 January 2020, https://www.express.co.uk/news/world/1230173/south-china-sea-china-outmatches-us-in-every-way-standoff-boils-spt
53 Gregory B. Poling, "The Conventional Wisdom on China's Island Bases Is Dangerously Wrong", War on the Rocks, 10 January 2020, https://warontherocks.com/2020/01/the-conventional-wisdom-on-chinas-island-bases-is-dangerously-wrong/
54 S. Paul. Kapur. "Dangerous Deterrent: Nuclear Weapons Proliferation and Conflict in South Asia". NUS Press, 2009.
55 Joy Mitra, "Will India's Response to Pulwama Deter Large-Scale Terror Attacks?", The Wire, 2018, https://thewire.in/security/balakot-airstrikes-india-pakistan-pulwama
56 Lieutenant General (Retd) Khalid Kidwai says, "Indians also should have understood that from there on, the rush to a nuclear crisis was but a few steps away and that there would be no choice for India, but to step back and look for face saving options involving international players, highlighting yet again the centrality of the core issue of Kashmir in South Asia". For complete speech see "Keynote Address and Discussion Session with Lieutenant General (Retd) Khalid Kidwai", International Institute of Strategic Studies, 2020, https://www.iiss.org/events/2020/02/7th-iiss-and-ciss-south-asian-strategic-stability-workshop. pg. 9.
57 "Lok Sabha Passes J&K Reorganisation Bill with 370 Votes for and 70 against It", 6 August 2019, https://zeenews.india.com/india/live-updates/live-updates-jammu-and-kashmir-reorganisation-bill-in-lok-sabha-for-scrapping-article-370-2225088
58 T. Negeen Pegahi, "The New India Versus the Nuclear Revolution: The Future of Crises Among Nuclear Powers", War on the Rocks, 12 June 2019, https://

warontherocks.com/2019/06/the-new-india-versus-the-nuclear-revolution-the-future-of-crises-among-nuclear-powers

59 Manoj Joshi, "The Wuhan Summit and the India–China Border Dispute", Observer Research Foundation, 26 June 2018, https://www.orfonline.org/research/41880-the-wuhan-summit-and-the-india-china-border-dispute/

60 "Current Stand-Off an Attempt by China to Change the Status Quo at Tri-Junction: Shivshankar Menon", The Wire, 2018, https://thewire.in/diplomacy/current-stand-off-attempt-china-change-status-quo-tri-junction-shivshankar-menon

61 Manoj Joshi, "The Wuhan Summit and the India–China Border Dispute", Observer Research Foundation, 26 June 2018, https://www.orfonline.org/research/41880-the-wuhan-summit-and-the-india-china-border-dispute/

62 Ibid.

63 In 2007 a number of Indian bunkers in the Batang La area were destroyed by China. See Manoj Joshi, "Doklam: To Start at the Very Beginning", ORF, 9 August 2017, https://www.orfonline.org/research/doklam-start-very-beginning/

64 "Prakash Menon & Anirudh Kanisetti "The Doklam Imbroglio, Takshashila Discussion Document, 2018-03", Takshashila, https://takshashila.org.in/takshashila-discussion-document-the-doklam-imbroglio/

65 "In Official Testimony to MPs, Government Revealed Full Story of Doklam", The Wire, 2018, https://thewire.in/diplomacy/doklam-parliamentary-standing-committee-india-china

66 "Prakash Menon & Anirudh Kanisetti, "The Doklam Imbroglio, Takshashila Discussion Document, 2018-03", Takshashila, https://takshashila.org.in/takshashila-discussion-document-the-doklam-imbroglio/

67 Col. Vinayak Bhat (retd. "Near Doklam, China Is Again Increasing Forces, Building Roads & Even a Possible Heliport", ThePrint, 2 April 2019, https://theprint.in/defence/near-doklam-china-is-again-increasing-forces-building-roads-even-a-possible-heliport/215512/

68 Chinese side maintained that the Indian side withdrew first. See "Foreign Ministry Spokesperson Hua Chunying's Regular Press Conference on 28 August 2017", Fmprc.gov.cn, 2017, https://www.fmprc.gov.cn/mfa_eng/xwfw_665399/s2510_665401/2511_665403/t1487932.shtml

69 Ibid, "QUESTION NO.*219 PERMANENT PRESENCE OF CHINESE TROOPS IN DOKLAM AREA", Ministry of External Affairs India, 2013, https://www.mea.gov.in/rajya-sabha.htm?dtl/29642/question+no219+permanent+presence+of+chinese+troops+in+doklam+area

70 "US Congresswoman Claims China Resumed Doklam Activities, India Says Status Quo Prevails", Hindustan Times, 26 July 2018, https://www.hindustantimes.com/world-news/china-quietly-resumed-activities-in-doklam-area-claims-us-official/story-n0TCFWccs419d9iTrD0JrI.html

71 Ibid.

72 Bo Zhou, "Doklam Stand-off with China: Will India Learn the Right Lessons or Pay as Nehru Did?", South China Morning Post, 8 January 2018, https://www.scmp.com/comment/insight-opinion/article/2127282/doklam-stand-china-will-india-learn-right-lessons-or-pay

73 Reuters, "Chinese Troops Armed with Iron Bars and Rocks Brawl with Indian Soldiers near Border", South China Morning Post, 15 August 2017, https://www.scmp.com/news/asia/diplomacy/article/2106954/chinese-and-indian-soldiers-said-involved-melee-indian-sources

74 "Scuffle between Indian and Chinese Troops in Ladakh as Two Incursion Bids by PLA Foiled", Economic Times, 15 August 2017, https://economictimes.indiatimes.com/news/defence/scuffle-between-indian-and-chinese-troops-in-ladakh-as-two-incursion-bids-by-pla-foiled/articleshow/60076876.cms

75 "Indian, Chinese Soldiers Get into Scuffle in Ladakh", Economic Times, 12 September 2019, https://economictimes.indiatimes.com/news/defence/indian-chinese-soldiers-get-into-scuffle-in-ladakh/articleshow/71091348.cms
76 Joel Wuthnow, "Doklam, One Year Later: China's Long Game in the Himalayas", War on the Rocks, 7 June 2018, https://warontherocks.com/2018/06/doklam-one-year-later-chinas-long-game-in-the-himalayas/
77 Reuters, "Chinese Troops Armed with Iron Bars and Rocks Brawl with Indian Soldiers near Border", South China Morning Post, 15 August 2017, https://www.scmp.com/news/asia/diplomacy/article/2106954/chinese-and-indian-soldiers-said-involved-melee-indian-sources
78 Detresfa_, and SIM TACK, "China-India Border Crisis Has Quietly Resulted in Victory for Beijing", The Drive, 23 May 2022, https://www.thedrive.com/the-war-zone/china-india-border-crisis-has-quietly-resulted-in-victory-for-beijing
79 Ibid.
80 Yogesh Joshi & Anit Mukherjee (2019) From denial to punishment: The security dilemma and changes in India's military strategy towards China, Asian Security, 15:1, 25–43
81 IANS, "Mountain Strike Corps Will Bridge Gaps in India's Defence Capabilites: Experts", Business Standard, 18 July 2013, https://www.business-standard.com/article/news-ians/mountain-strike-corps-will-bridge-gaps-in-india-s-defence-capabilites-experts-113071800654_1.html. And By SUJAN DUTTA, "Tezpur to Host Four Sukhoi 30MKI- Squadron Base Means Air Force Moving Powerful Assets near China", Telegraph India, June 2009, https://www.telegraphindia.com/india/tezpur-to-host-four-sukhoi-30mki-squadron-base-means-air-force-moving-powerful-assets-near-china/cid/617651 And "With an Eye on China, India Decides to Station First Squadron of Rafale Fighter Jets in Bengal", India TV, 7 January 2017, https://www.indiatvnews.com/news/india-with-an-eye-on-china-india-decides-to-station-first-squadron-of-rafale-fighter-jets-in-bengal-364015. And Ajai Shukla, "In a First, Indian Tank Brigades to Defend China Border", Business Standard, 16 September 2012, https://www.business-standard.com/article/economy-policy/in-a-first-indian-tank-brigades-to-defend-china-border-112091702018_1.html
82 Brad Roberts, "U.S. Nuclear Weapons in the Twenty-First Century", 2016, https://carnegieendowment.org/files/Transcript17.pdf. pg. 5
83 This point is brought out in Snyder's analysis of extended deterrence over two alliance territories that are not equally valuable. One is more valuable and equally so to the deterrer and aggressor in comparison to the other that has a lesser value. In such cases the threat of higher escalatory responses is less credible owing to the fact that the stakes involved are very low compared to the cost of fighting. See Glenn H. Snyder. "Deterrence and Defense". Princeton University Press, 1961. doi:10.2307/j.ctt183pj49. pg. 22, 23.
84 Michael Fitzsimmons, "The False Allure of Escalation Dominance", War on the Rocks, 16 November 2017, https://warontherocks.com/2017/11/false-allure-escalation-dominance/. And Brad Roberts, "U.S. Nuclear Weapons in the Twenty-First Century", 2016, https://carnegieendowment.org/files/Transcript17.pdf. pg. 6.
85 Austin Long. "Deterrence -- From Cold War to Long War: Lessons from Six Decades of RAND Research". RAND Corporation, 2008, https://www.rand.org/pubs/monographs/MG636.html. pg. 11–13.
86 For Stability-Instability paradox see Krepon Michael, Rodney Jones & Ziad Haider, "Escalation Control and the Nuclear Option in South Asia", Accessed 1 May 2020, https://www.stimson.org/wp-content/files/file-attachments/Escalation%20Control%20FINAL_0.pdf. Pg 2 and Glenn H. Snyder. "Deterrence and Defense". Princeton University Press, 1961. doi:10.2307/j.ctt183pj49. pg. 226.
87 Michael J. Mazarr. "Understanding Deterrence". RAND Corporation, 2018, https://www.rand.org/pubs/perspectives/PE295.html. pg. 10.

88 Ibid, John J. Mearsheimer. "Conventional Deterrence". Cornell University Press, 1983. www.jstor.org/stable/10.7591/j.cttlrv61v2. pg. 53.
89 Kier A. Lieber & Daryl G. Press, "Coercive Nuclear Campaigns in the 21st Century; Understanding Adversary Incentives and Options for Nuclear Escalation", Georgetown University, 2013.
90 Kier A. Lieber & Daryl G. Press. "Preventing Escalation during Conventional Wars". 2015. pg. 3.
91 Lieber and Press include both implicit and explicit threats of nuclear escalation in their analysis. Footnoted in Ibid, pg. 8.
92 Stimson Center, "Contesting the Indo-Pacific: Military Technology and the Shifting Offense-Defense Balance", YouTube Video, 2020, https://www.youtube.com/watch?v=7utHXtSJcaw&feature=youtu.be. 0.00-22.00
93 There is the possibility of the advertent overlap of nuclear and conventional systems, this aspect is discussed in the last chapter.
94 Jay Ross, "Time to Terminate Escalate to De-Escalate — It's Escalation Control", War on the Rocks, April 24, 2018, https://warontherocks.com/2018/04/time-to-terminate-escalate-to-de-escalateits-escalation-control/
95 Ibid.
96 T. Negeen Pegahi, "The New India Versus the Nuclear Revolution: The Future of Crises Among Nuclear Powers", War on the Rocks, 12 June 2019, https://warontherocks.com/2019/06/the-new-india-versus-the-nuclear-revolution-the-future-of-crises-among-nuclear-powers/
97 We are not concerned here with higher alert status of nuclear weapons which could lead to inadvertent escalation; that point will be dealt with in later chapters.
98 Gerald C. Brown, "Deterrence, Norms, and the Uncomfortable Realities of a New Nuclear Age", War on the Rocks, 20 April 2020, https://warontherocks.com/2020/04/deterrence-norms-and-the-uncomfortable-realities-of-a-new-nuclear-age/ and "Statement of Robert Scher Assistant Secretary of Defence for Strategy, Plans, and Capabilities before the Senate Armed Services Subcommittee on Strategic Forces", 2016, https://www.armed-services.senate.gov/imo/media/doc/Scher_02-09-16.pdf. pg. 2.
99 Al Mauroni, "Tearing Down the Nuclear Firewall", War on the Rocks, 15 October 2019, https://warontherocks.com/2019/10/tearing-down-the-nuclear-firewall/
100 Quoted in Amy F. Woolf, "Nonstrategic Nuclear Weapons", 2017, https://fas.org/sgp/crs/nuke/RL32572.pdf. pg. 8.
101 See Amy F. Woolf, "Nonstrategic Nuclear Weapons", Congressional Research Service, 2017, https://fas.org/sgp/crs/nuke/RL32572.pdf
102 For instance, a B61-12 nuclear bomb has yield of 50 kilotons, but a guided tail kit affords it more accuracy to meet the targeting capability of the B61-7, a strategic bomb with a much larger maximum yield of 360 kilotons. See Hans Kristensen, "Non-Strategic Nuclear Weapons", 2012, https://fas.org/_docs/Non_Strategic_Nuclear_Weapons.pdf. pg. 24.
103 Al Mauroni, "Tearing Down the Nuclear Firewall", War on the Rocks, 15 October 2019, https://warontherocks.com/2019/10/tearing-down-the-nuclear-firewall/
104 Kahn lists this at the Rung 19 of his escalation ladder. Kahn Herman. "On Escalation: Metaphors and Scenarios". Vol. 1. Transaction Publishers, 2009.
105 Michael Fitzsimmons, "The False Allure of Escalation Dominance", War on the Rocks, 16 November 2017, https://warontherocks.com/2017/11/false-allure-escalation-dominance/
106 There is an opportunity cost with induction of TNWs: one, they take spending away from conventional forces towards maintenance of an arsenal which can only be used in very rare circumstances, and second if the yield is low enough to allow battlefield use, then the military damage they might inflict could also be low. Finally,

they take away fissile material from the strategic arsenal, whose size is more critical to general and strategic deterrence, survival, and the limitation of conflict.

107 Gerald C. Brown, "Deterrence, Norms, and the Uncomfortable Realities of a New Nuclear Age", War on the Rocks, 20 April 2020, https://warontherocks.com/2020/04/deterrence-norms-and-the-uncomfortable-realities-of-a-new-nuclear-age/

108 Michael Fitzsimmons, "The False Allure of Escalation Dominance", War on the Rocks, 16 November 2017, https://warontherocks.com/2017/11/false-allure-escalation-dominance/

109 Balance of terror works on a different form of power, that of inflicting severe punishment and to deter by threat of punishment as opposed to the power to contest the control of territory. Snyder refers to this in the context of nuclear weapons, but this is equally applicable for higher-order conventional responses like missile strikes and the use of long-range fighter jets. See Glenn H. Snyder. "Deterrence and Defense". Princeton University Press, 1961. doi:10.2307/j.ctt183pj49. pg. 41.

110 The probability of annihilation at the conventional level relative to the nuclear level is low and stretched out over time, so states are more risk-prone to use conventional means. See John J. Mearsheimer. "Conventional Deterrence". Cornell University Press, 1983. www.jstor.org/stable/10.7591/j.ctt1rv61v2. pg. 23.

111 Austin Long. "Deterrence -- From Cold War to Long War: Lessons from Six Decades of RAND Research". RAND Corporation, 2008, https://www.rand.org/pubs/monographs/MG636.html. pg. 45.

112 John P. Rose. "The Evolution of U.S. Army Nuclear Doctrine, 1945-1980". Westview Press, 1980. pg. 85.

113 Michael Krepon, Rodney W. Jones & Ziad Haider. "Escalation Control and the Nuclear Option in South Asia". Henry L. Stimson Center, 2004. pg. 125.

114 Ibid, pg. 128.

115 Glenn H. Snyder. "Deterrence and Defense". Princeton University Press, 1961. doi:10.2307/j.ctt183pj49. pg. 140.

116 Ibid, pg. 127.

117 Michael Krepon, Rodney W. Jones & Ziad Haider. "Escalation Control and the Nuclear Option in South Asia". Henry L. Stimson Center, 2004. pg. 155, 156.

118 "Limited Strategic War", edited by Klauss Knorr & Thornton Read, Frederick A. Praeger, 1962. pg 11–12.

119 John J. Mearsheimer. "Conventional Deterrence". Cornell University Press, 1983. www.jstor.org/stable/10.7591/j.ctt1rv61v2. pg. 29.

120 "Joint Publication 3-72 Nuclear Operations". Pentagon, 11 June 2019, pg. 33.

121 Particularly greater dispersion on the battlefield can greatly increase the number of TNWs required for achieving the same battlefield effect. See Nayyar A & Zia Mian. "Pakistan Security Research Unit (PSRU) Brief Number 61: The Limited Military Utility of Pakistan's Battlefield Use of Nuclear Weapons in Response to Large Scale Indian Conventional Attack", 2010. https://sgs.princeton.edu/sites/default/files/2020-01/nayyar-mian-2010.pdf. pg. 6–9.

122 Vincent A. Manzo & Aaron R. Miles, "The Logic of Integrating Conventional and Nuclear Planning", Arms Control Association, 2016, https://www.armscontrol.org/act/2016-10/features/logic-integrating-conventional-nuclear-planning#notes. Also see Keir A. Lieber & Daryl G. Press. "Preventing Escalation during Conventional Wars". 2015. pg. 38.

123 "The Logic of Integrating Conventional and Nuclear Planning", Arms Control Association, 2016, https://www.armscontrol.org/act/2016-10/features/logic-integrating-conventional-nuclear-planning#notes

124 Keir A. Lieber & Daryl G. Press. "Preventing Escalation during Conventional Wars". 2015. pg. 38.

125 Vincent A. Manzo & Aaron R. Miles, "The Logic of Integrating Conventional and Nuclear Planning | Arms Control Association", Armscontrol.org, 2016, https://

www.armscontrol.org/act/2016-10/features/logic-integrating-conventional-nuclear-planning#notes

126 First use is not first strike.

127 "Statement of Robert Scher Assistant Secretary of Defence for Strategy, Plans, and Capabilities before the Senate Armed Services Subcommittee on Strategic Forces", 2016, https://www.armed-services.senate.gov/imo/media/doc/Scher_02-09-16.pdf. pg. 3.

128 Tami Davis Biddle, "Coercion Theory: A Basic Introduction for Practitioners", Texas National Security Review, 20 February 2020, https://tnsr.org/2020/02/coercion-theory-a-basic-introduction-for-practitioners/

129 "Statement of Robert Scher Assistant Secretary of Defence for Strategy, Plans, and Capabilities before the Senate Armed Services Subcommittee on Strategic Forces", 2016, https://www.armed-services.senate.gov/imo/media/doc/Scher_02-09-16.pdf. pg.3.

130 "Deterrence Stability Escalation Control In South Asia", edited by Michael Krepon & Julia Thompson, 2013, https://www.stimson.org/wp-content/files/file-attachments/Deterrence_Stability_Dec_2013_web_1.pdf. pg. 70.

131 Ibid, pg. 71.

132 John J. Mearsheimer (1989) Assessing the conventional balance: The 3:1 rule and its critics, International Security, 13:4, 54–89. DOI: 10.2307/2538780

133 Since a limited space/objective is to be defended this results in a higher force to space ratio.

134 John J. Mearsheimer. "Conventional Deterrence". Cornell University Press, 1983, www.jstor.org/stable/10.7591/j.cttlrv61v2. pg. 44.

135 In such a situation the any attack that penetrates the defence is certain to be decisive. See Mearsheimer John J. "Conventional Deterrence". Cornell University Press, 1983. Accessed 23 April 2020, www.jstor.org/stable/10.7591/j.cttlrv61v2. pg. 44.

136 "Vijay Prahar: Army Men Practise Fighting in 'Nuclear Weapon Environment'", The Economic Times, 7 May 2018, https://economictimes.indiatimes.com/news/defence/vijay-prahar-army-men-practise-fighting-in-nuclear-weapon-environment/articleshow/64067961.cms; Exercise Poorna Vijay, see "India's Limited War Doctrine: The Structural Factor", Manohar Parrikar Institute for Defence Studies and Analyses, 2016, https://idsa.in/monograph/IndiasLimitedWarDoctrine_aahmed. pg. 34; Shatrujeet Exercise (2016) see Frank O'Donnell, "Reconsidering Minimum Deterrence in South Asia: Indian Responses to Pakistan's Tactical Nuclear Weapons", Contemporary Security Policy, 2016, DOI: 10.1080/13523260.2016.1236470. pg. 9–10.

137 Dinshaw Mistry, "Complexity of Deterrence among New Nuclear States: The India- Pakistan Case" in Complexity of Deterrence: Strategy in the Global Age (University of Chicago Press, 2019) doi: 10.7208/9780226650043-010. pg. 196.

138 Ibid.

139 Amina Afzal, "What's in a Name?" South Asian Voices, 9 December 2013, https://southasianvoices.org/whats-in-a-name/

140 These cities are roughly 50 km, 190 km, 90 km, and 130 km respectively from the nearest points on the border with India. See A Nayyar and Zia Mian, "Pakistan Security Research Unit (PSRU) Brief Number 61: The Limited Military Utility of Pakistan's Battlefield Use of Nuclear Weapons in Response to Large Scale Indian Conventional Attack", 2010, https://sgs.princeton.edu/sites/default/files/2020-01/nayyar-mian-2010.pdf. pg. 5.

141 Cited in Feroz Khan, "Calhoun: The NPS Institutional Archive Faculty and Researcher Publications Faculty and Researcher Publications Battlefield Nuclear Weapons and Deterrence Strategies: Phase III", Accessed 26 April 2020, https://core.ac.uk/download/pdf/36737488.pdf. pg. 16.

142 Frank O'Donnell & Debalina Ghoshal (2018) Managing Indian deterrence: pressures on credible minimum deterrence and nuclear policy options, The Nonproliferation Review, 25:5–6, 419–436, DOI: 10.1080/10736700.2019.1565187, 9–10.
143 "The Cabinet Committee on Security Reviews Operationalization of India's Nuclear Doctrine", Ministry of External Affairs, 2013, https://www.mea.gov.in/press-releases.htm?dtl/20131/The+Cabinet+Committee+on+Security+Reviews+perationalization+of+Indias+Nuclear+Doctrine
144 Sandeep Unnithan, "Does the Air Force Have a Plan to Take out Pakistan's Tactical Nuclear Weapons?", India Today, 6 October 2017, https://www.indiatoday.in/india/story/indian-air-force-pakistan-tactical-nuclear-weapons-nasr-bs-dhanoa-1058876-2017-10-06
145 Olya Oliker. "Russia's Nuclear Doctrine; What We Know, What We Don't, and What That Means". Center for Science and International Security , 5 May 2016, https://www.csis.org/analysis/russia%E2%80%99s-nuclear-doctrine
146 Michael Kofman, "It's Time to Talk About A2/AD: Rethinking the Russian Military Challenge", War on the Rocks, 5 September 2019, https://warontherocks.com/2019/09/its-time-to-talk-about-a2-ad-rethinking-the-russian-military-challenge/
147 "Special Report: Nuclear Posture Review - 2018", U.S. Department of Defense, 2018, https://dod.defense.gov/News/SpecialReports/2018NuclearPostureReview.aspx
148 Vipin Narang, "The Discrimination Problem: Why Putting Low-Yield Nuclear Weapons on Submarines Is So Dangerous", War on the Rocks, 8 February 2018, https://warontherocks.com/2018/02/discrimination-problem-putting-low-yield-nuclear-weapons-submarines-dangerous/
149 Russia is said to have around 2000 non-strategic nuclear weapons. See Hans M. Kristensen & Matt Korda (2019) Tactical nuclear weapons, Bulletin of the Atomic Scientists, 75:5, 252–261, DOI: 10.1080/00963402.2019.1654273
150 Amy F. Woolf, "Nonstrategic Nuclear Weapons", Congressional Research Service, 2017, https://fas.org/sgp/crs/nuke/RL32572.pdf. pg. 12.
151 Heginbotham Eric, Michael Chase, Jacob Heim, Bonny Lin, Mark Cozad, Lyle Morris, Christopher Twomey, et al. "China's Evolving Nuclear Deterrent Major Drivers and Issues for the United States". RAND Corporation, 2017, https://www.rand.org/content/dam/rand/pubs/research_reports/RR1600/RR1628/RAND_RR1628.pdf. pg. 15–20.
152 Richard D. Fisher, Jr, "Missile Investments Are Needed to Meet China's Nuclear Challenge", The Diplomat, 25 April 2020, https://thediplomat.com/2020/04/missile-investments-are-needed-to-meet-chinas-nuclear-challenge/
153 Gregory Kulacki, "Would China Use Nuclear Weapons First in a War With the United States?", The Diplomat, 27 April 2020, https://thediplomat.com/2020/04/would-china-use-nuclear-weapons-first-in-a-war-with-the-united-states/
154 Heginbotham Eric, Michael Chase, Jacob Heim, Bonny Lin, Mark Cozad, Lyle Morris, Christopher Twomey, et al. "China's Evolving Nuclear Deterrent Major Drivers and Issues for the United States". RAND Corporation, 2017, https://www.rand.org/content/dam/rand/pubs/research_reports/RR1600/RR1628/RAND_RR1628.pdf. pg. 28–31.
155 Cited in Heginbotham Eric, Michael Chase, Jacob Heim, Bonny Lin, Mark Cozad, Lyle Morris, Christopher Twomey, et al. "China's Evolving Nuclear Deterrent Major Drivers and Issues for the United States". RAND Corporation, 2017, https://www.rand.org/content/dam/rand/pubs/research_reports/RR1600/RR1628/RAND_RR1628.pdf. pg. 67.
156 Tong Zhao, "Conventional Counterforce Strike and Damage Limitation", https://igcc.ucsd.edu/_files/PPNT/NuclearBriefing2presentation1.pdf
157 China's conventional missile forces have the capability to inflict damage on U.S. air base operations within roughly 1,500 km of Chinese territory. See Nixon, Michael, Forrest Morgan, Jacob Heim, Jeff Hagen, Sheng Li, Jeffrey Engstrom, Martin

Libicki, et al., "US-China Military Scorecard: Forces Geography and the evolving balance of power 1996-2017", https://www.rand.org/content/dam/rand/pubs/research_reports/RR300/RR392/RAND_RR392.pdf. pg. 45–70.

158 Eric Heginbotham, Michael Chase, Jacob Heim, Bonny Lin, Mark Cozad, Lyle Morris, Christopher Twomey, et al. "China's Evolving Nuclear Deterrent Major Drivers and Issues for the United States". RAND Corporation, 2017, https://www.rand.org/content/dam/rand/pubs/research_reports/RR1600/RR1628/RAND_RR1628.pdf. pg. 159.

159 Frank O'Donnell. "India and Nuclear Asia: Forces, Doctrine, and Dangers". Belfer Center for Science and International Affairs, 2019, https://www.belfercenter.org/publication/india-and-nuclear-asia-forces-doctrine-and-dangers. pg. 81–87.

160 Ibid, pg. 110.

161 Ajai Shukla. "China Has Captured 60 Sq Km of Indian Land!". Rediff.com, June 8, 2020, https://www.rediff.com/news/special/china-has-captured-60-sq-km-of-indian-land/20200608.htm

162 Christopher Clary & Narang Vipin, "India's Pangong Pickle: New Delhi's Options After Its Clash with China", War on the Rocks, July 2, 2020, https://warontherocks.com/2020/07/indias-pangong-pickle-new-delhis-options-after-its-clash-with-china/

163 Suhasini Haider. "Ladakh Face-off | Chinese Build-up Started in May: MEA", The Hindu, June 25, 2020, https://www.thehindu.com/news/national/ladakh-face-off-china-amassing-troops-armaments-at-lac-since-early-may-says-india/article31917068.ece

164 Aidan Milliff, "Tension High, Altitude Higher: Logistical and Physiological Constraints on the Indo-Chinese Border", War on the Rocks, June 8, 2020, https://warontherocks.com/2020/06/tension-high-altitude-higher-logistical-and-physiological-constraints-on-the-indo-chinese-border/

165 Bharat Karnad, "An out and out Army Fiasco", Security Wise, June 15, 2020, https://bharatkarnad.com/2020/06/15/an-out-and-out-army-fiasco/#:~:text=Inattentive%20to%20satellite%20intel%2C%20lacking,has%20to%20explain%20this%20fiasco

166 Sushant Singh, "Explained: What Happened to the Mountain Strike Corps?", The Indian Express, June 19, 2020, https://indianexpress.com/article/explained/explained-what-happened-to-the-mountain-strike-corps-6465221/

167 Frank O'Donnell & K. Bollfrass Alexander. "The Strategic Postures of China and India: A Visual Guide". Belfer Center for Science and International Affairs, March 2020, https://www.belfercenter.org/publication/strategic-postures-china-and-india-visual-guide

168 Global Times, Twitter, June 24, 2020, https://twitter.com/GlobalTimesBiz/status/1275628988879130624

2

FROM KILL CHAIN TO KILL WEB

Militaries depend heavily on completing their decision cycles with efficiency and swifter than their adversaries. This decision cycle is a function of objectives, doctrines, and technological ability and acts like overhead costs for militaries; the shorter the better. In the third nuclear age, the rise of new and emerging technologies impacts these decision loops in multiple ways and forces militaries to exploit certain capabilities, deal with vulnerabilities, and change the way they want to structure their forces and organization, to operate faster in their decision cycle. These emerging technologies are likely to play a pivotal role in shaping the decision cycle and deciding if mutual military vulnerabilities or capabilities are creating opportunistic military motivations and advantages.[1] This chapter looks at the role of emerging technology in reshaping the decision cycles of militaries in the third nuclear age and the ramifications of this altered decision cycle on conflict incentives and deterrence at the operational or tactical levels.

The chapter starts by revisiting the targeting methods in Operation Desert Storm in 1991 and the Iraq War in 2003 and then moves onto the concept of net-centricity in military operations, the meaning and evolution of the "kill chain" concept. Subsequently, it lays out the contours of the next stage of evolution in the kill chain which is the "kill web". Traditionally decision loops have been completed by a cycle which is based on the concept of a kill chain that is domain centric. The kill web is an evolution of the kill chain into a multidimensional, more automated, and more densely networked version which is seamlessly cross-domain in nature. The central goal of the kill web concept is to break down the domain barriers to coagulate the different domains into one singular battlespace where effects in one domain can be delivered from any other domain with equal battlespace consequences. This implies that the kill web concept places equal importance for land, air, sea, space, cyber, and electronic domains where both kinetic and non-kinetic effects can achieve battlefield impact. Finally, the chapter

DOI: 10.4324/9781003365976-3

looks at the manner in which kill web integrates or aligns an instrumentalized conventional strategy and how conventional-nuclear integration features prominently in the kill web. The kill web's effectiveness here is moderated by instrumentalized conventional strategy for controlling escalation.

Gulf Wars, evolutions, and revolutions

It was first in the 1991 Iraq War that coalition forces executed a form of "parallel warfare" that concretized a combination of stealth, precision, and long-range or stand-off capability as the preferred template of warfare.[2] The outcome was an evolution of an initially proposed targeting plan for the air force called the "instant thunder" that centred around a concept called the "five rings model".[3] This five rings model essentially described a state in terms of concentric rings, which had a strategic category of targets like leadership at the innermost ring, and subsequently key production centres, infrastructure, population, and fielded forces in the outer rings which essentially protected the inner rings. The strategy followed was that of "inside-out warfare", meaning an inner ring had higher priority compared to an outer ring in terms of the importance attached to the category of targets in a campaign which was to be led by the aerial forces. In practice, the actual phase 1 of the four-phase Operation Desert Storm saw the application of both air power and ground forces, and the additional targets that were taken out during phase 1 of the operation were an addendum to the initially proposed instant thunder plan upon which these additional targets were superimposed to cater for the needs of battlespace interdiction.[4]

The instant thunder concept involved simultaneous attacks upon strategic centres of gravity throughout the entire theatre of war.[5] The concept was used in Operation Desert Storm where it established a template that revolved around the use of large-scale and highly intense conventional air campaign as a prelude to the ground-offensive[6] along with extensive use and effectiveness of the Tomahawk land attack missiles. Coalition forces launched some 297 Tomahawk missiles,[7] 64 per cent of which were launched within the first two days of the beginning of the operation.[8] In total, 17,000 precision-guided munitions and 2,10,000 unguided munitions were used. Finally, high-value targets were struck "systematically without the need for a separate sustained preliminary battle for air supremacy".[9] The idea was to use air and missile power holistically to achieve a "strategic effect" by attacking "centres of gravity" and to render the leadership ineffective or incapacitated to pursue its political or military goals.[10] This operational orientation towards "strategic effect" was different from simply achieving an "interdiction",[11] as the latter could be used to achieve an effect only on the battlespace front whereas the former focused on coercing and pressurizing the leadership directly.[12]

Under the instant thunder concept, concurrent attacks by stealth, stand-off, and precision-guided weapons did not guarantee an adversary's defeat; they had to focus on the leadership to meet the political objectives.[13] These improved technologies that were available to the military forces had to be employed in a combined manner towards an objective. As shown in Tables 2.1 and 2.2, the objective was to render the leadership ineffective of pursuing its politico-military objectives rather than engage in a force-on-force contest.[14] This was supposed to achieve an instantaneous rather than a graduated cumulative effect. Critics, however, have argued that in practice the strategic effects were never truly achieved nor practical and that airpower deployed towards "strategic interdiction" had a substantial impact in achieving the military objectives.[15] Proponents of the five

TABLE 2.1 The growth of targets: from instant thunder to phase I

Target Set	*Instant Thunder*	*Phase I*
Leadership	5	33
Command control and communication	19	56
Electricity	10	17
Oil	6	12
Nuclear biological and chemical weapons	8	23
Military support/production and storage	15	73
Railroads and bridges	3	33
Strategic air defence	10	56
Airfields	7	31
Scuds	–	48
Republican guard	–	37
Naval ports	1	17
Breaching	–	–
Surface-to-air missiles	–	45
Total	84	481

TABLE 2.2 Coalition strikes by target category for Operation Desert Storm

Target	*Strikes*	*%*
Leadership	260	0.7
Command, control, and communications	580	1.7
Electric power	280	0.8
Oil	540	1.5
Nuclear, biological, and chemical	990	2.8
Military industry (production/storage)	970	2.8
Lines of communication	1,170	3.3
Surface-to-air missiles	1,370	3.9
Scuds	1,460	4.2
Iraqi Air Defence System (KARI)	630	1.8
Airfields	2,990	8.5
Naval targets	370	1.1
Iraqi ground forces	23,430	66.90
Total	35,040	100

Source: Cohen et al., The Gulf War Air Power Survey: Effects and Effectiveness, pg 1.

rings model argued that overall greater precision did make it possible to execute targets simultaneously, but it was the targeting plan borne out of greater precision, which was more impactful. The use of missiles and air power concurrently against strategic target categories within the first few days of the conflict was the outcome in terms of revolution in military affairs (RMA).

System of systems and the 2003 Iraq War

This targeting plan crystallized into the concept of "system of systems" first introduced by Admiral William Owens in 1996 which was the antecedent to the "network-centric warfare" concept.[16] He predicted an RMA via a dense synergy in three general categories of ISR/sensors: intelligence collection, surveillance, and reconnaissance; C4: command, control, communications, computer applications, and intelligence processing; and finally precision force.[17] This essentially addressed the fundamental issue of joint warfare and combined arms[18] and at the heart of this integration was the requirement of generating a "common operating picture" for all forces.[19] In the "system of systems" concept, this common operating picture was predicated on "dominant battlespace knowledge" that relied on networks to reduce "fog and friction of war" and bestow real-time awareness and status of forces.

The progress on this was notable within a decade. During the 1991 campaign, ground assault was preceded by a long bombing campaign and it typically took three days from the target being assigned to a plane to an eventual hit.[20] In contrast, during the 2003 Iraq War, soldiers on the ground were able to hand-off targets to air assets that could then hit the target in a total of mere ten minutes.[21] Relatively, United States and its allies during their global "war on terror" did operate with perhaps the best common operating picture.[22] Yet this was not a result of a well-structured architecture for joint warfare or a "system of systems".[23] Many of the key C4I systems were still heavily "service centric" and linkages were conjured up or workarounds evolved as combat progressed.[24]

The scale of the ISR effort was larger in comparison to the 1991 Iraq War, 42,000 battlespace images, 2,400 mission hours of SIGINT coverage, 32,000 hours of mission video, and 1,700 hours of moving target indicator coverage.[25] But this was not without problems, both technical and flow related. For instance, the manoeuvre formations were without direct access to satellite primary imagery, due to prioritization of requirements at the command level.[26] The information and data that were being made available were not tailored to the mission requirement[27] and this burdened the limited bandwidth. Finally, despite the ISR intelligence and data that was available, the responsibility to figure intentions or plans of the enemy commander had to fall on analysts and HUMINT.[28] Interoperability and variation in communication equipment often led to one unit receiving communication over different networks and radios, and units had to switch between different kinds of equipment due to incompatibility issues.[29,30]

The United States used nearly ten times more precision-guided weapons in 2003 relative to unguided munitions during the 1991 campaign.[31] In the 1991 war, only a small number of aircrafts had necessary avionics for acquiring targets, illuminating them for launching laser-guided bombs or acquiring target coordinates from the ground for precision strike missions, but in the 2003 Iraq War, all aircrafts were able to launch laser and other precision-guided bombs.[32] This allowed for a wider set of targets to be hit very early in the conflict.[33]

Unmanned aerial vehicles (UAVs) like Predator and Global Hawk were used for surveillance and targeting.[34] Tactical UAVs played an important role in circumventing the tendency to prioritize the provision of imagery to higher echelons of command rather than the actual warfighting elements.[35,36] By making available actionable intelligence, they met the requirement of tactically relevant information for high operational tempo and time sensitivity.[37]

Overall, though network-centric operations were being implemented to a degree at the higher echelons, net-centric capability was not realized at the level of a division, battalion, or below. Many of the effects that were generated were because of the platform and number-oriented method of conducting operations rather than the net-centricity of the overall effort. The kill chain was still sequential, could only act against stationary or pre-planned targets and the emphasis was on amassing firepower.[38] The method or template remained the same compared to 1991 and was only better executed in terms of timing and capabilities thanks to an evolution in technology particularly at the level of the platform.

Kill chain to the kill web

At the heart of NCW was the concept of the "kill chain" which in the words of Christian Brose allowed militaries to do three specific things on the battlespace: "understanding about what is happening", "make a decision about what to do", and finally "take action that creates an effect to achieve an objective". Future warfare, however, could witness militaries "close the kill chain" more effectively and swiftly, while hindering the inability of the adversary to do the same.[39] This efficiency would need a higher degree of NCW capabilities which must get intelligence directly to the warfighting element.[40] Until the kill chain era, this intelligence could be efficiently and reliably delivered only within the same domain. But in the third nuclear age, cross-domain transmission of intelligence and information is progressively becoming a norm in achieving true NCW capabilities in what is referred to as the kill web.

At this point, it is important to note the concept of domains that have traditionally come to define the battlespace. These traditionally included air, land, maritime, space, and cyberspace.[41] Though some others have considered electromagnetic spectrum as a separate additional domain of warfare mindful of its importance.[42] Historically armies battled armies, navies against navies, but that changed with the advent of air power that brought firepower from the skies to effect outcomes on land and in the naval theatre in World War II. As time progressed, modern militaries developed single-domain capabilities that could inflict damage in another domain. This bearing of effect, however, did not imply that domain boundaries had been eradicated, rather that they could now be utilized towards achieving a common purpose in the overall battlespace. As technological progress made it possible to better coordinate, synchronize actions in each domain towards a combined battlespace effect, it yielded concepts like "system of systems", "network-centric warfare", "network-centric operations", "network of networks", etc.[43]

Yet, as brought out in the earlier sections throughout Operation Desert Storm and the Iraq War 2003, there was only an evolutionary improvement and not a revolutionary change in breaking down these domain barriers. Consequently, the "service-centric" nature of the linkages survived even as the currency of bringing to bear a networked or system of systems in the battlespace gained importance. Each domain had its own observe-orient-decide-act (OODA) loop,[44] and this stove-piping reflected in the limits to jointness that could be conjured up. For instance, a kill chain that was to be executed across two domains had to be necessarily sequential, implying that the kill chain of one domain had to be completed before the kill chain of the other domain could begin.

Ray Walderman has the most lucid explanation of how the overall OODA loop operates,

> "the Observe segment is done by reconnaissance aircraft and satellites, or ground and ship-based radar and Signals Intelligence (SIGINT) systems. The Orient segment processes the raw data to gain insight into the situation. The Decide segment establishes what tactical moves will be made, at what time, with what resources. The Act segment takes longer to move troops and ships into a fight than it does for aircraft and missiles. As it turns out, the Act phase is simply the combined kill chain of all the domains and their stovepipes."

According to Walderman, this combined kill chain follows the 5F model of "find, fix, fire, finish, feedback" and the rate-determining step in executing this model of the combined kill chain is the slowest step in the chain.[45] This was a consequence of the fact that domain-specific kill chains operated in a sequential manner and all of whom operated at different speeds.[46] Army devices that could operate on land could not exchange information with Navy systems, while Navy systems could not exchange information with Air Force systems.[47] Installation of common equipment with standard protocols, common network architecture, and data structures was as much a technical problem as it was an organizational turf problem.[48] In addition, different platforms whether on land, air, or in the sea use different voltages to power their equipment, and radio systems are very environment constricted. An air force electronic system needs to operate in a high-altitude low-pressure environment with tolerance for extreme temperatures; Army systems on the other hand need to be resilient against "shock and vibration, wider temperature extremes, sand, dust, dirt, ice, snow, rain, and humidity".[49] Finally, some platforms could only send data while others could only receive data.[50] Beyond the challenge of what was doable was the challenge of how and why it should be done. For example, should stealth platforms send data at all when it could comprise their stealth? Beyond that a second issue during target engagement is whether the "tracking information" being sent over the network is of "engagement quality", meaning is it consistently fast and good enough to allow a direct hit by a shooter. For this purpose, the targeting information must be dynamically updated and sent over the network with a velocity that allows another networked platform to effectively engage the target. While a simple way to do this would be to just connect everything together, it is not practical because it inevitably burdens the network with unnecessary data as not all nodes in the network need to communicate with every other node and the information that needs to be sent over the network also has to be filtered to what is necessary rather than sending everything over a finite bandwidth.

The current set of concepts like "cross-domain", "multi-domain", and "sensor-to-shooter" therefore originate and find their value in addressing these impediments of single-domain sequential kill chains by moving towards a concept of "kill webs" or "kill networks".[51] A kill web allows for both cross-domain detection and cross-domain engagement, and it is in this battlespace that the impact, kinetic, or non-kinetic is delivered by the kill web.[52]

Brose calls this kill web a battle network composed of "sensors" that collect information, "shooters" that act on that information. Though it is essentially the "sharing of the information" at the level of the network that is invaluable.[53] In the kill web, sensors are dispersed in the battlespace, but particularly low-cost low-tech platforms (usually unmanned systems) add high density[54] that was hitherto unavailable at operational-level depths in the battlefield. The high-tech platforms on the other hand have sensor fusion and precision armament-delivering capabilities to take advantage of the penetrating view generated by these low-cost expendable numerous remote-operated or autonomous platforms. The process of converting raw data into an engagement quality track (referred to as "sensor fusion") is dependent on combining inputs and then using high processing power and artificial algorithms to get targeting data. The kill web therefore exploits the "high-low mix" of platforms with sensor fusion capabilities to the hilt to augment situational awareness.[55] The backbone of this web architecture is obviously dependent on processing and moving information fast to the node where it is needed. Irrespective of the domain a platform is in, the network requires C4ISR on those platforms whether it's sonar, radar, signals intelligence (SIGINT), sensors, electronic warfare (EW), cyber warfare (CW), COMM (networked military communications), and administrative systems. These platforms of large or small real-estate, limited or large numbers and weapon carrying or ISR oriented come together to form the kill web. Kill web is therefore an accumulated capability but distributed in capacity diminishing the importance of large complex platforms that can only be procured or fielded in limited numbers.

Finally, this sensor network in the kill web is critically dependent on satellites for the primary functions of "find, identify, fix, track, fire, finish, and feedback" along with battlefield communications.[56] Contrary to the Cold War understanding when space as a domain was not concerned with conventional warfighting within the kill web, space as other domains has become an inseparable part of the singular conventional battlespace. The rising ease and lower cost of launching and building satellites have enabled militaries to take advantage of space-based capabilities to close the kill web by provisioning both high-speed communications and high-resolution imagery/intelligence thus potentially enabling time-sensitive targeting.[57,58] The ramification of this is felt in the conventional-nuclear intersection which will be discussed in more detail in the last chapter.

Situational awareness, intelligence, and space

Kill web is characterized by a very "high density of sensors" that includes "radar, sonar, infrared, acoustic (sound), electromagnetic or signals intelligence or electronic intelligence (SIGINT and ELINT), electro-optical or Imaging intelligence (IMINT), motion detectors to cyber warfare systems are focused on the target area by satellites, ground vehicles, troops, planes, helicopters, drones, ships, and submarines".[59] "These various types of sensors collect information in the battlespace for any kind of signature the adversary units might emit (find), process it into targeting intelligence (Identify), and share those targets on the network" to generate engagement quality information. This has also been made possible by rapid improvement in sensor quality. Sensor quality not only improves data collection methods but also improves the value of data itself.[60] For example, in some cases like Light Detection and Ranging (LIDAR) sensors, accuracy of the map could be improved with a trade-off in terms of lower data collection rate.[61] While Synthetic Aperture Radar (SAR) can now detect moving targets and compute both their direction and speed.[62] To add to this sensor quality introduction of high number of unmanned assets means a very high density of sensors in the tactical battlespace area.

Kill web therefore cuts out dependence on single platforms, and instead a network of large number of unmanned and numerous decentralized LEO satellites are used to acquire targets.[63] Unlike JSTARS, which becomes critical in the absence of a network and therefore prime targets for adversary long-range armament, webs are more resilient. They therefore better exploit enhanced visibility along with an information and communication sharing network that can pass on attributes related to the target(s) within a completely visible battlespace, like location coordinates, velocity, etc. This allows for engagement quality information to be passed and processed across the entire web or network within an appreciably small time interval ranging from minutes to seconds or even mini-seconds.[64] This is an invaluable capability that can adapt itself to varied emerging threats in real time and act on them using the most efficient means at their disposal, that is, pairing the right shooter with the right sensor across domains. Situational awareness is particularly more effective in the air and on the sea than on land because the radar returns and heat emissions are more easily detected and distinguishable on the sea and in the air from background clutter and second because land offers terrain obstacles where forces can afford to take cover.[65] Even on land, concealment has become tough over time as even under dark and natural cover conditions the electronic signatures generated by the communication and data-sharing systems of ground troops can be spotted by opponents.[66]

Situational awareness of the kill web is where the coalescence of the domains begins. In particular, space domain is fundamental to conduct of conventional warfare for all advanced powers, and to a great degree, information supremacy and thereby situational awareness are critically dependent on space-based assets which therefore transitions into an area of operational vulnerability.[67]

> "Communication satellites broadcast content and relay data, remote sensing platforms collect images and signals for intelligence and commercial uses that can be combined with other data. Satellite computers in orbit connect to ground stations and to each other via radio data links."

Overall space and cyberspace combine to host much of the information infrastructure for command, control, communication, computation, intelligence, surveillance, and reconnaissance.[68] One of the key roles of space-based systems is to transmit data over the horizon something territorial systems can't do.[69] It enables "missile defense, targeting of mobile forces at distance, military operations through positioning, navigation, and timing (PNT) services and secure communications".[70]

While an ICS strategy in principle is unlikely to look for gaining first-strike advantages in space, dependency and criticality of space to the ICS effort do degrade first-strike stability in space. This instability has also increased on account of greater target indistinguishability where C4ISR satellites related to conventional capabilities can be targeted distinctly from nuclear C2 infrastructure[71] and without comprising a state's ability to detect nuclear missile launches[72] or targeting satellites for commercial usage. This has repercussions for ICS, as though adversary space systems might not be targeted during the intra-war deterrence phase where the objective is escalation control, the same may not be applicable in the peacetime coercion phase where a first strike could lead to critical mobilization advantages and allow a tactical win. This is particularly enabled by developments of certain kinetic and non-kinetic means that do not produce debris or have the ability to temporarily blind satellites.[73] There are many ways in which space-based situational awareness can be degraded; for example, both Russia and China have been developing laser systems that can shoot down satellites in LEO.[74] Other means include jamming, cyberspace operations, and co-orbital attacks.[75] Kinetic or non-kinetic hits against ground-based satellite stations can also degrade space-based situational awareness. Such capabilities force space as a domain to become an integral part of the kill web particularly favourably disposed towards peacetime coercion phase of the ICS.[76] Against nuclear weapon states that don't exhibit space resilience or lack in redundancy or where segregation between conventional and nuclear C2 is not possible, such attacks may not align with the peacetime coercion stage of ICS.

Density, network, and autonomy

The greater density of sensors both on the platform and across the network implies that the volume of data that is generated by the kill web in the network is greater than anything realized in a kill chain and instead of centralizing large volume of information, storage and processing is done at the edge of the network.[77] A corollary of this evolution is that the availability of processing power at the edge of the network results in an overall decentralized architecture of the network where the processing power would reside. The algorithm at the level of the platform itself has the ability to analyze raw data for pertinent information which is then relayed to C2 nodes nearer to the front. After a second level of collection at the C2 level, processed data with the relevant information is sent over the kill web.

The automated processing ability at the edge ensures the network moves less volume and the most pertinent data from the edge to the rest of the web.[78] Finally, this ensures that the web can quickly decide and route the information to the warfighting element that is in need of the information and decision support.[79]

A more important consequence of the introduction of greater computing power and thereby data processing prowess and a better learning outcome for algorithms allows the consumption of large volume of data to identify patterns of behaviour that hitherto could not be done simply because the required resources or computing power were not available. As artificial intelligence trains on the data generated from the dense sensors in the web, acts on it, and feedback is generated, the web can evolve and adapt itself to get exponentially faster and better in understanding adversary decision-making as the war progresses.[80] These advances and their application may still be a few years away from ripeness, but they can only bring greater efficacy and power to the kill web particularly at the edge.

It also means that in a kill web, some sort of automation must exist to assist or even takeover human decision-making, given the throughput of information. Historically states have limited themselves to fighting their adversaries in the domain in which the crisis has occurred and only later expanded into other domains.[81] With the kill web, this choice might come under pressure sooner rather than later. The choice will become even harder as networks of manned and unmanned assets across land, sea, and air come together in a kill web. The cognitive capacity required to take control of the battlefield assets could be beyond humans and the complexity has to be managed with computers.[82] As the battle networks become greater, autonomy in the network could ensure that speed at which engagement happens at the frontier surpasses the human decision-making timelines. This would force an even greater unmanned platform mix and autonomous command and control at the lower echelons.[83] This has implications for command and control of the kill web that will be discussed later in the text.

Attrition, precision, distribution, and range

Precision, distribution, and attrition are the last mile attributes of the kill web. No matter how capable a single platform may be, it is always restricted in space and the kind of sensors and communication systems that can be placed on it. In a networked kill web, these capabilities are distributed across the network, and a collection of decentralized networks that are reconfigurable makes the overall kill web more reliable and resilient to attacks.[84] This allows for distributed form of capabilities in terms of sensing, kinetic, electronic, or cyber effects that could tilt the scales in favor of a force that is agile and can bring in quantity due to a low cost of individual units and can field the same capability in a distributed manner while it has the capacity to take attrition. This quality of attrition combines with precision, range, and stealth to make the kill web extremely effective.

A case in point is the latest Azerbaijan-Armenia conflict that evinced some of these important trends in warfare that highlight how distribution aided with precision could enhance the efficacy of the kill web of future in terms of effecting both a strategic outcome or interdiction which are very relevant to ICS. Azerbaijan first started its campaign on 27 September 2020 in the Nagorno-Karabakh conflict, but took losses only at the edge of the force envelope; in contrast, the Armenian forces took losses both on the frontier and in-depth.[85] This was largely due to Azerbaijan using a combination of suicide drones and long-range fires to degrade the Armenian ability to supply frontline troops by targeting their assembly areas, command posts, logistics, and manoeuvre elements.[86] Loitering munitions with their small size evade radar detection and could be employed in large numbers due to their low cost to inundate targets of high value such as SAM sites.[87] In addition, it used Harop loitering munition against Armenian S-300 surface-to-air missile (SAM) sites.[88]

While the use of Turkish Bayraktar TB2 UAV with

> "reasonable loiter time and guided munitions allowed Azeri forces to launch hunter-killer raids into the Armenian rear to find and interdict Armenian forces. They have also provided intelligence, surveillance and reconnaissance feeds enabling greater situational awareness and the accurate employment of artillery and multiple launch rocket systems that lack guidance, turning cheap conventional capabilities into deadly accurate long-range fires systems."

Each of these low-cost drones are in fact cheaper in comparison with manned assets or reusable UAVs by an order of magnitude.[89] In essence, they allowed range and accuracy to be brought on to the field by allowing conventional systems like artillery with even dumb munitions to calibrate and deliver pinpoint fire at the target.

On the other hand, accuracy and long-range aspects are also introduced by using missiles. Azerbaijan for instance used a Long-Range Artillery (LORA) Weapon System developed by Israel Aerospace Industries with an accuracy of 10m circular error probable (CEP) and guidance by global navigation satellite system (GNSS) and inertial navigation system (INS) and supported by TV for terminal guidance[90] to reliably hit targets of value. Earlier in a similar instance on 8 January 2020, Iranian missile attacks on the American bases in al-Asad and Erbil, Iraq also evinced a CEP of 5–10 m.[91] Unlike the previous generations of ballistic missiles that had a CEP of 150–900 m, this level of precision allows the ability to strike military logistics hubs and critical infrastructure at range and effectively.[92] Azerbaijan was thus able to inflict damage via precision strikes at operational depth without a heavy modernized air force at a fraction of the defence budget of the world's major powers.[93] As a result, progressively Armenian frontline resistance weakened allowing Azerbaijan to make territorial gains.

Kill web can also take a high level of attrition on account of two factors: one, the losses are primarily suffered on account of frontline unmanned assets, autonomous or remotely controlled, and second, this frontline attrition is more in material and not in men, which can be replenished much faster if own centres of gravity are protected to maintain the operational tempo. In addition, kill web exudes range and precision. The ability of the web to deliver kinetic effects using short- to medium-range armament, loiter munitions, or kamikaze drones and armed drones particularly at range and with precision is realized in a greater ability to inflict damage remotely and therefore convey deterrence more effectively or use the web more purposefully with the conventional military strategy. China for instance particularly places a great degree of confidence in its missile arsenal for taking out targets close to its periphery[94]; it extends this strategy by extending the "home advantage" with conventional missiles that can hit accurately at greater ranges and thus pushing this home advantage envelope. Systems like DF-21 and DF-26 could potentially utilize the situational awareness afforded by the kill web to target military bases and US carrier battle groups (CBGs) in the near seas and they do so by combining enhanced awareness with precision strike capability.[95] This lends greater efficacy to the web in executing its tactical missions resulting in deterrence at the operational level of the theatre while also forcing adversaries to revaluate their force structures and their operational procedures. This operational deterrence is the intra-war deterrence that comes to the fore once the peacetime coercion objective is fulfilled.

This potential of the web can be explained with a boxer analogy. Kill web acts like a boxer that has great endurance such that it can take multiple punches from a more skillful and better trained pugilist and still be able to counter-attack due to relatively infinite endurance. The other boxer (something less than a kill web) even if more skillful cannot counter this endurance because sooner or later the web will inflict critical damage that will degrade its fighting potential. Greater density also allows the kill web to detect and see with greater efficacy and the throughput in the network allows instant kills against anything that can be detected. The density of sensors generated due to both manned and unmanned platforms in the kill web allows it greater capability as well as capacity to take more attrition and the web in fact thrives on the network's ability to generate quality information on engagement throughout the network by using multiple platforms rather than simply more technologically capable platforms. This essentially means that a military that may be able to field higher capable platforms in smaller numbers may not necessarily do well against a military that may field slightly inferior platforms but that are numerous and networked. In addition, platform-related stealth attributes and stealth as a concept may have a short life span, specially once the density of sensors reaches a critical mass in the web to complete the common battlespace picture. In the Armenia-Azerbaijan conflict, there are instances where Armenian troops attempted concealing, used decoys, and tried digging in, and yet at least two of these positions were struck by Azerbaijani forces.[96] The ability of autonomous networks to fuse data from multiple and varied types of sensors to identify the entity in question could reduce stealth to a non-attribute.[97]

A point to note here is that improvements in guidance technologies, better picture of the battlespace and the targets, and greater sensor density (or distribution) in the web is an ongoing evolutionary trend since the Gulf War. This allows for a precision-strike based targeting strategy that can be effectively towards adversary command and control elements for achieving an instantaneous strategic effect as envisaged by Warden or towards interdiction in the battlespace. Even if critics believed the former was not achievable or practical during the Gulf War, the evolution of the kill chain into the kill web with ongoing improvements in precision technology make Warden's conception of instant strategic effect look increasingly feasible.

Kill web and instrumentalized conventional strategy

States are likely to attempt integrating kill web within the overall contours of an instrumentalized conventional strategy. It may in fact cater to all "instrumentalized" uses of conventional forces and military strategy. The web's ability to integrate kinetic and non-kinetic effects towards an outcome renders it compatible with a conventional instrument that is geared towards peacetime coercion and can seamlessly transition to serve the role of escalation management or intra-war deterrence. The operationalization of the evolving web by multiple military forces is a likely indicator of how kill web mated with ICS is likely to emerge as a doctrine and could be used to secure strategic outcomes potentially without engaging in large-scale conventional battles.

Cyberattacks and electronic warfare could be used to effect peacetime operations that could allow coercion of adversaries to achieve strategic outcomes below the threshold of large-scale military hostilities. The effects generated here are purposeful to utilize synergy on land, cyber, and electronic realms towards peacetime coercion. These non-kinetic means add to the potential of the kill web.

While Chinese doctrinal thinking espouses an effects-based kill web concept articulated in terms of Anti-Access/Area Denial (A2/AD).[98] The weapon systems that implement this A2/AD strategy are dependent on the C4ISR systems that compose the kill web and intend to degrade the US forces which are critically dependent on space and cyber assets.[99] China has placed dense integrated missile and air defence systems in the acquired island territories and high-power jammers to degrade communications and to detect approaching US aircraft at longer ranges.[100] This integration strives to take out bases within the theatre while degrading the ability of the United States to project force by taking out the logistics chain before they could move into the theatre from greater distances and is geared more towards escalation control.

ICS strategy derives strategic outcomes by revisiting "interdiction" as a principle of warfare and kill web makes it a feasible option. This is like an effects-based operations approach for strategic outcomes, except it is limited in geographical terms, is sequential, and achieves a limited objective over time in the form of implementing fait accompli(s). Conventional means under ICS are instrumentalized both towards peacetime coercion and intra-war deterrence, but ICS places a greater emphasis on peacetime coercion and a sequential move towards escalation control. Nuclear-armed states can use their evolving kill web capabilities to support ICS strategy both for peacetime coercion or intra-war deterrence stages. The Galwan clash in 2020 between India and China amidst the pandemic merits attention here. In Galwan, Chinese were able to take advantage of the fact that Indian military could not conduct a scheduled exercise[101] due to the coronavirus pandemic and its ISR capability was particularly weak to detect an intrusion.[102] Chinese thus were able to mobilize earlier in strength and impose a fait accompli. It is not difficult to imagine how similar conditions could be imposed on an adversary by blinding or taking out satellites using non-kinetic means like directed energy weapons or other soft-kill methods, while the transport systems like railways and airbases are disabled by a cyber-attack in peacetime to allow own forces time to deploy or accomplish a fait accompli. Post the Galwan clash between India and China, Chinese quickly reverted to intra-war deterrence posture. Though unmanned capability aspect was not brought out in a big way in this event, China did display swarms of drones delivering food to frontline soldiers, a capability that could in effect also be used to bomb Indian frontline positions, especially if these were network-centric swarms that could not be defended against by an adversary without specialized defences. While, Chinese missile capabilities were always a factor in the background.

The mobilization advantages derived in the peacetime coercion stage of the ICS are in alignment with the "interdiction" method of targeting, where the idea was to derive advantage at the forward edge of the battlespace since the military objectives were limited. Instead of trying to achieve a strategic effect by going for a targeting method as envisaged by Warden with a healthy percentage of munitions allocated directly against high-level command and control nodes, the kill web capabilities used by ICS at this stage strive to maintain the greatest segregation between conventional-nuclear spectrums and operational and tactical levels of the conflict to convey limited goals.

Kill web is equally efficient in conveying intra-war deterrence between NWS. As expounded in the earlier chapter, once peacetime coercion is implemented the intra-war deterrence or escalation control becomes the primary objective of an ICS, the targeting strategy evolves from interdiction to achieving strategic effect geared towards intra-war or operational-level deterrence against the adversary where the kill web already has the necessary capability due to expansion of the web's ability to generate a better situational awareness better communication or achieve greater force concentration in shorter time. In this phase, the kill web targeting method resembles Warden's conception within the theatre of confrontation and conveys greater tolerance towards escalation. The distinction between tactical and operational levels is erased to a great degree in this phase as the targeting strategy changes.

There are a variety of ways in which the web can operationalize effective escalation control once peacetime fait accompli has been operationalized in line with ICS. Drones for example could combine with traditional air assets and create advantages for air forces against land forces by taking away the limited advantages they may have in terrain or operational factors, as well as reduce the dependency of air forces on competent and friendly troops on ground for tactical air coordination for conducting strikes. Unlike previous decades when air forces could hit hard, halt offensives but not conduct offensive operations with a decisive impact without friendly ground forces,[103] unmanned systems are likely to remove these impediments. Earlier forces operating on land could afford to disperse in the absence of a credible ground threat, the arrival of unmanned systems that can perform ISR roles both on land and air could makes this an untenable option.[104] By making use of air warfare more likely for offensives, they render the military cost-benefit calculus of the recipient of ICS unfavourable and make the actor more amenable to persisting with the newly established status quo than taking the risk of restoring it.

Escalation control or intra-war deterrence can also be implemented by calibrating in terms of the targets that are attacked or the domain-centric chain that is operationalized in the kill chain. For example, in 2019 when India retaliated with air strikes against Pakistan over a militant attack in Pulwama, a similar effect for example could have been achieved with a ground-based operation using special forces or artillery strikes or even a cyber strike that could produce similar physical effects. But the visibility and the political outcome of an air strike could not have been exhibited by other domains in the kill web. In its response, Pakistan chose a matching aerial response to the Indian strike to re-establish intra-war deterrence. Escalation is therefore not just a function of the target but also of the targeting method and this is relevant to ICS. Both states managed escalation control in this iteration by limiting the duration and iteration of engagement of forces while matching their responses in the aerial domain. This highlighted a key link between ICS and the kill web in which technologies that are more malleable in terms of proportionality are more likely to aid intra-war deterrence strategies. While technologies that are more cross-domain in nature are more likely to aid peacetime coercion strategies. Given that ICS is inherently a sequence of peacetime coercion followed by intra-war deterrence stage, it is likely that cross-domain operationalization of the kill web will be followed by single-domain kill chains that aid in escalation control and establish domain or method-related thresholds that actors can adhere to in a crisis.

The pace of this transformation of the conventional instrument from serving peacetime coercion to its transformation into an instrument of intra-war deterrence and escalation control is a key feature of the third nuclear age that is enabled and supported by the kill web. This is one of the key reasons why seemingly small crises seem to progress faster and appear to acquire scale or intensity or both and escalation seems more probable. This favourable disposition towards quickly moving from accomplishment of the fait accompli to intra-war deterrence is fundamental to the way instrumentalized conventional strategy operates and a key reason why crises seem to progress faster in intensity or scale.

Kinetic and non-kinetic paths in the kill web

Both kinetic and non-kinetic means are available within the kill web and can deliver effects at longer ranges. The combination of the kinetic and non-kinetic effects in the kill web could give way to a battlespace that is enlarged.[105] This battlespace can be shaped in-depth or at the front, the hit can be scored at both civilian and military targets, and the hit can be scored in peacetime or during active hostilities. In line with the instrumentalized conventional strategy, the effects of non-kinetic means may tend to precede the kinetic kill chain to ensure that the fait accompli is actualized. Once the fait accompli is actualized, both the kinetic and non-kinetic components of the kill web in combination may serve as an instrument of intra-war deterrence, and if this intra-war deterrence breaks down, then the loss of control over the adhering to threshold could take a linear or non-linear path towards the conventional-nuclear transition point.

Though the kill web attempts to effectively breakdown domain barriers for the participants in peacetime coercion both the initiator and recipient of ICS, the perception of escalation prospects is domain-centric and therefore domain boundaries still represent strong or weak thresholds depending on the theatre and the perception of the actors involved in ICS. This observation has also been witnessed in tabletop games[106] where for instance aerial means are almost always seen as more disposed towards escalation. While non-kinetic means like cyber operations may not have the same visible effect as that of kinetic means despite having the same destructive or mission kill effect, rendering them more prone to being used for peacetime coercion.[107] A state could use an overt air strike as a means of retaliating to a terrorist strike or it could conduct a covert cyber kill against command and control node.[108] But the political visibility and the deterrent and threshold value against the adversary of these two methods will be different.

Depending on the theatre in which the conflict has begun, the use of certain forces could be perceived as less escalatory; for instance, if a conflict began in the maritime domain, the use of naval forces would be seen as less escalatory and would not cross perceived thresholds. Whereas in a conflict use of non-kinetic capabilities like electronic warfare could be seen as an asymmetric advantage in that they don't engage in a force-on-force contest yet render systems in land, air, or at sea ineffective by achieving a "mission kill" or manipulating targeting data.[109] Analysts have noted that such non-kinetic capabilities are increasingly dispersed across military hierarchy as seen in the case of Russian military where electronic warfare units are available at brigade and division levels.[110] Electronic warfare for instance has been used in Syria against AC-130J gunships by Russian forces by jamming their communication links and GPS receivers rendering them unable to coordinate their attacks with other manned and unmanned aircrafts or ground-based tactical air controllers.[111]

These are considered asymmetric capabilities relative to kinetic hits. In terms of capability, however, both conventional kinetic means and non-kinetic means are equally capable in inflicting damage even with strategic consequences. Blurring the distinction between conventional and nuclear realms, essentially in a way where this distinction is merely reduced to target selection rather than the mode of attack. Particularly in the non-kinetic realm due to the bidirectional flow of data, cyber warfare has a role to play and has the potential to take out the warfighting ecosystem that sits at the intersection of conventional and nuclear capabilities and operations.

Although some analysts have noted that cyber weapons need to be tailored to the target set or type and that they required persistent access which is difficult if not impossible,[112] use of cyberspace within the kill web is considered intrusive and its potential for escalation is generally considered to be higher given they can target both conventional and nuclear command and control nodes. This escalatory potential is moderated by ICS strategy that leads to the operationalization of the kill web, so the intrusive and potentially escalatory effect of the use of cyberspace capabilities for offensive purposes is likely to be mitigated by the fact that the likely use is for interdiction as the method of targeting rather than achieving an instantaneous strategic effect not suited for peacetime coercion. This is relatively easily done with targets in the logistics space than say in command and control nodes because there is complexity in judging and segregating conventional and nuclear command and control targets.[113] Cyberspace capabilities that penetrate sovereign systems might be seen as highly provocative acts by the receiver especially if those space-based or other systems serve both conventional and nuclear C2s.[114] As most times the line between intrusion for surveillance and an actual attack is only the intent of the actor[115] and some contend that this could create "use or lose" pressures.

Escalation prospects

Escalation prospects due to the web can be captured in two characteristics peculiar to the web: strategy and entanglement. Historically it has been observed that technology is neither necessary nor a sufficient condition for escalation.[116] Overall, the kill web explained above aligns with the view that technology acts as an intervening variable rather than an independent variable driving inadvertent or nuclear escalation.[117] By its very nature, since the kill web is built around a larger number of networked but cheap systems, losing a few to achieve an operational or tactical outcome in the battlespace is a more preferred choice than not taking active measures. Therefore, the kill web naturally aligns better with compellent policies than passive deterrence strategies. The one with the larger network of cheaper individual systems in the kill web is likely to be more inclined and favourably disposed towards offensive and first-strike options.[118]

As a matter of strategy, ICS is generally intolerant towards escalation in the first phase as it seeks to use adequate and minimal force for a calculated threshold and incremental limited objectives to further strengthen the kill web's ability to reinforce itself in line with ICS. As the web transitions at the intra-war deterrence stage, its targeting approach mimics instant thunder to achieve a strategic effect with respect to the theatre of conflict. Under ICS, kill web seeks to achieve a mutually reinforcing position for the web to convey operational-level deterrence or control over escalation. However, as the web pivots from accomplishing the peacetime coercion to intra-war deterrence, the speed of the transition can introduce escalation into the conflict especially if an executed fait accompli results in active hostilities. Overall, as ICS decides the method and intent of targeting when the kill web is operationalized, the web itself should provoke escalation only when other factors in the realm of politics and strategy are also in alignment to increase prospects of escalation.[119]

But if that is not the case and the web starts leading strategy, then the web itself could also be a source of escalation. James Johnson for instance writes in the context of a conflict between the US and China that as both actors are "dependent on advanced C4ISR capabilities to support their offense dominant capabilities", this could "increase the incentives to strike first, lower the threshold for the use of military force, cause 'use or lose' situations, and compounded by misperception, cause deterrence failure and inadvertent cross-domain warfare". Again, this scenario is an outcome of technology acting as an independent variable rather than being led by a strategy or policy. Particularly states for instance might see a "first strike" advantage in degrading the adversary kill web while their own kill web is still intact. Although this point is likely to be reached only when intra-war deterrence strategies fail or if there is misreading of thresholds for escalation control. The fact that the kill web can suffer a greater level of attrition does allow for crisis stability; however, if this attrition lets situational awareness capabilities degrade, then crisis stability could degrade as the damage inflicted on the web's ability to perceive the battlespace correctly is an order of magnitude higher damage than the attrition of warfighting assets at the edge. The loss of situational awareness could degrade the web's ability to function as a network in supporting both conventional and nuclear operations. For instance, some sensors meant for use in conventional contingencies and supporting conventional operations can be used for tracking the nuclear mobile missiles of NWS with a small arsenal.[120] These include High Altitude Long Endurance (HALE) drones, constellation of small satellites, etc., which coupled with air power, cruise missiles, and other conventional strike assets, enable counterforce doctrines,[121] thereby in effect degrading first-strike stability and creating "use or lose" pressures for nuclear powers with relatively small arsenals.

This "use-or-lose" situation could potentially lower the threshold for use of nuclear weapons in the conflict. States that do not have the ability to suffer attrition because they lack redundancy or resilience are likely to generate first mover's advantage over their nuclear rivals. This might lead to a reverse ramification leading vulnerable states to develop military doctrines that are based on "first-use" with ambiguous red lines. However, as Catelyn Talmadge has convincingly argued, though technology may create the perception of first mover's advantage, in the end technology can only enable or accelerate risks originating in state policies or military doctrines.[122] As discussed earlier, ICS favours swift transition from peacetime coercion to intra-war deterrence. Given the precision kill web exhibits, it makes it possible to target C2 nodes as the most efficient way towards taking out over-the-horizon radars or integrated defence systems.[123] But the onus to escalate here falls upon the state on the receiving end of the ICS. Yet again inadvertent ramifications due to entanglement can degrade crisis stability and hasten crisis progression at the intra-war deterrence stage of ICS. This entanglement includes "dual-use systems, co-mingling of nuclear and conventional forces and their support infrastructures, and non-nuclear threats to nuclear weapons and their associated C3I systems," which is discussed in greater detail in the later chapters.

Conclusion

The kill web is surely an abstract and a futuristic concept for now but one that should merit greater attention as the battlespace gets more digitized. As is evident, the many developments and innovations in terms of new technologies, platforms, armaments, or artificial intelligence do not in themselves constitute a revolution unless they fit into an overall puzzle which is the kill web capable of generating more efficient effects both in the kinetic and non-kinetic domains towards a politico-military goal. This kill web is a further improvement on the single domain-centric kill chain and allows for cross-domain sensing and shooting. In addition, the kill web allows for much greater level of situational awareness through a greater density of sensors in a decentralized combination of networks. These networks are autonomous at the edge and perform military operations through a network of manned and unmanned assets that are capable of suffering attrition even as they deliver effects through kinetic or non-kinetic means. Kill web does exhibit conventional-nuclear intersection while it gives the conventional military instrument the ability to create a strategic effect or achieve interdiction on the battlespace.

If technology were to independently lead strategy, then the lethality and the precision of the kill web are naturally more prone towards first strike and going after strategic targets like C2 nodes that make the prospect of both advertent and inadvertent escalation very likely. However, the kill web is more likely to be used in an instrumentalized conventional strategy which in effect aids a method of targeting that allows the conventional capability to sequentially support peacetime coercion and then pivot rapidly towards intra-war deterrence or escalation control. There is a constant tension between the capability of the kill web that could be used for strategic effects as conceptualized by Warden and the limited war objectives between nuclear adversaries, but ICS moderates the technological capability of the web towards interdiction to support peacetime coercion. Kill web does allow for swift transition of the military instrument from peacetime coercion to intra-war deterrence once the military gains are made which is a key feature of the third nuclear age and the reason why crises at the lower end of the spectrum may appear to progress faster in intensity and scale. This kill web-enabled swift transition in the roles served by the conventional military instrument introduces operational-level deterrence into tactical-level situations blurring the difference between tactical and operation-level confrontations. This is not a surprising conclusion given that this manuscript started with conventional and nuclear realms as part of one continuous conflict spectrum. The next chapters will look at the impact of unmanned units in the kill web, the role of cyber, space, and hypersonics, and the implications for the conventional-nuclear intersection.

Notes

1 Todd S. Sechser, Neil Narang & Caitlin Talmadge (2019) Emerging technologies and strategic stability in peacetime, crisis, and war, Journal of Strategic Studies, 42:6, 727–735, DOI: 10.1080/01402390.2019.1626725, 731
2 Cited in John Andreas Olsen. “Strategic Air Power in Desert Storm”. Routledge, 2003. pg. 233.
3 Barry Watts, “The Effectiveness of Coalition Air Power Against Iraq's Nuclear Weapons Program & Modified ‘Scud’ Missiles during Desert Storm”, OSD Counterforce Workshop, RDA Logicon Arlington, USA, 19 July 1995.
4 Dietz Matt. “Toward a More Nuanced View of Airpower and Operation Desert Storm” War on the Rocks, January 6, 2021. https://warontherocks.com/2021/01/toward-a-more-nuanced-view-of-airpower-and-operation-desert-storm/

5 Deptula, "Parallel Warfare", pp. 127–156 cited in John Andreas Olsen, Routledge, 2003. pg. 149.
6 John Andreas Olsen. "Strategic Air Power in Desert Storm", Routledge, 2003, and Thomas A.K. & Eliot A.C., "Revolution in Warfare? Air Power in the Persian Gulf", Foreign Affairs, 1996, August, https://www.foreignaffairs.com/reviews/capsule-review/1996-07-01/revolution-warfare-air-power-persian-gulf
7 Of these, 282 were successfully launched. See John Andreas Olsen. "Strategic Air Power in Desert Storm". Routledge, 2003. pg. 242 and Norman Friedman. "Desert Victory: The War for Kuwait". Naval Institute Press, 1991. pg. 339.
8 Ibid.
9 John Andreas Olsen. "Strategic Air Power in Desert Storm". Routledge, 2003. pg. 244.
10 Ibid, pg. 129.
11 An example of air-based interdiction is the US air campaign against the North Vietnamese Army between 1968 and 1970 known as Operation Commando Hunt, when despite performance issues interdiction created hurdles for movement of NVA troops and in effect deterred NVA troops from initiating their invasion for several months and had the eventual outcome of forcing NVA to change its primary military objective away from Saigon to demilitarized zones. For more on this, see Gartzke Erik & Jon R. Lindsay, eds. "Cross-domain Deterrence: Strategy in an Era of Complexity". Oxford University Press, 2019. pg. 153–155.
12 John Andreas Olsen. "Strategic Air Power in Desert Storm". Routledge, 2003. pg. 129.
13 Ibid, pg. 273.
14 Ibid, pg. 105–106.
15 See Dietz Matt. "Toward a More Nuanced View of Airpower and Operation Desert Storm", War on the Rocks, January 6, 2021, https://warontherocks.com/2021/01/toward-a-more-nuanced-view-of-airpower-and-operation-desert-storm/ and Pietrucha, Mike, "The Five-Ring Circus: How Airpower Enthusiasts Forgot about Interdiction", War on the Rocks, September 29, 2015, https://warontherocks.com/2015/09/the-five-ring-circus-how-airpower-enthusiasts-forgot-about-interdiction/
16 "Network-Centric Warfare: A New War Theory Emerging in the Information Age", Unrevealed Files, February 1, 2022, https://www.unrevealedfiles.com/network-centric-warfare-a-new-war-theory-emerging-in-the-information-age/
17 Collectively referred to as C4ISR. See Owens William A. "The Emerging US System-of-Systems". No. 63. National Defense University, Institute for National Strategic Studies, 1996.
18 Anthony H. Cordesman. "The Iraq War: Strategy, Tactics and Military Lessons". Natraj Publishers, 2009. pg. 180, 181.
19 Ibid, pg. 182.
20 Noah Shachtman. "How Technology Almost Lost the War: In Iraq, the Critical Networks Are Social — Not Electronic", Wired, November 27, 2007. https://www.wired.com/2007/11/ff-futurewar/
21 Ibid.
22 Anthony H. Cordesman. "The Iraq War: Strategy, Tactics and Military Lessons". Natraj Publishers, 2009. pg. 182.
23 Cited in Anthony H. Cordesman. "The Iraq War: Strategy, Tactics and Military Lessons". Natraj Publishers, 2009. pg. 182.
24 Ibid.
25 Ibid, pg. 184.
26 Ibid, pg. 185.
27 Ibid, pg. 188
28 Ibid, pg. 189.
29 Ibid, pg. 198.

30 Ibid, pg. 198–199.
31 Ibid, pg. 218. Overall the coalition fired 19,948 precision-guided munitions versus 8644 and 955 cruise missiles against 300 in the Desert Storm.
32 Ibid, pg. 276.
33 Ibid, pg. 259.
34 Ibid, pg. 286.
35 Ibid, pg. 313.
36 Ibid, pg. 315.
37 Ibid, pg. 312–313.
38 Brose Christian. "The Kill Chain: Defending America in the Future of High-Tech Warfare". Hachette UK, 2020. pg. 68.
39 Ibid, pg. 73.
40 Cited in Anthony H. Cordesman. "The Iraq War: Strategy, Tactics and Military Lessons". Natraj Publishers, 2009. pg. 184.
41 Feickert Andrew. "Defense Primer: Army Multi-Domain Operations (MDO)". Congressional Research Service, January 16, 2020. https://fas.org/sgp/crs/natsec/IF11409.pdf.
42 Alderman Ray, "Integrating Warfare Models with the Kill Chain", Military Embedded Systems, April 27, 2018, https://militaryembedded.com/radar-ew/sensors/integrating-warfare-models-with-the-kill-chain
43 Alderman Ray, "Origins of the Kill Web", Military Embedded Systems, August 29, 2020, https://militaryembedded.com/radar-ew/sigint/origins-of-the-kill-web.
44 OODA Loop is a four-step decision-making framework. The cycle begins with an observation, which leads a participant to orient on possible options, then decide on an appropriate course of action, and finally act on that decision. At that point, the results are observed, and the cycle begins again. In this version, success is a function of cycling through the OODA Loop faster than one's opponent. For more on this, see Alastair Luft, "The OODA Loop and the Half Beat", The Strategy Bridge, March 17, 2020, https://thestrategybridge.org/the-bridge/2020/3/17/the-ooda-loop-and-the-half-beat
45 Alderman Ray, "Integrating Warfare Models with the Kill Chain", Military Embedded Systems, April 27, 2018, https://militaryembedded.com/radar-ew/sensors/integrating-warfare-models-with-the-kill-chain
46 Alderman Ray, "Transitioning from the Kill Chain to the Kill Web", Military Embedded Systems, May 30, 2018, https://militaryembedded.com/comms/communications/transitioning-from-the-kill-chain-to-the-kill-web
47 Ibid.
48 Ibid.
49 Ibid.
50 Alderman writes, for instance, that while F-35 stealth fighter was designed from ground up for sharing battlespace awareness with platforms like drones F-15s, F-16s, and F-22s, could only receive targeting data and not share it with other platforms.
51 Dr. Greg Kuperman, "Adapting Cross Domain Kill-Webs", DARPA, 2022, https://www.darpa.mil/program/adapting-cross-domain-kill-webs
52 Throughout this text, the term battlespace has been used instead of battlefield to signify that unlike the traditional battlefield limited to the physical domains of air, land, sea, and space, battlespace involves electromagnetic, cyber and cyberspace, and cyber-electronic domains as well.
53 Brose Christian. "The Kill Chain: Defending America in the Future of High-Tech Warfare". Hachette UK, 2020. pg. 391–392.
54 This high-density augmentation of the network follows the Metcalfe's Law. Efficacy of the network is directly proportional to the number of noded connections in the network. In a network with N nodes, this would amount to a maximum N(N-1) connections.

55 Alderman Ray, "Fighter Jets and the Kill Web", Military Embedded Systems, June 28, 2018, https://militaryembedded.com/avionics/computers/fighter-jets-and-the-kill-web
56 Alderman Ray, "Satellites and the Kill Web", Military Embedded Systems, November 25, 2019, https://militaryembedded.com/comms/satellites/satellites-and-the-kill-web
57 Brose Christian. "The Kill Chain: Defending America in the Future of High-Tech Warfare". Hachette UK, 2020. pg. 309.
58 Ibid.
59 Alderman Ray, "How the Kill Web Manipulates Time and Space", Military Embedded Systems, March 31, 2020, https://militaryembedded.com/comms/communications/how-the-kill-web-manipulates-time-and-space
60 Hersman Rebecca, Reja Younis, Bryce Farabaugh, Bethany Goldblum & Andrew Reddie, "Under the Nuclear Shadow: Situational Awareness Technology and Crisis Decision Making", Center for Strategic and International Studies, March 18, 2020, https://csis-website-prod.s3.amazonaws.com/s3fs-public/publication/200318_UnderNucearShadow_FullReport_WEB.pdf?VJm_nrx2bVVeByYH38yx8YkDvvr1QZVW. p14
61 Cited in Ibid, pg. 14.
62 Ibid.
63 Brose Christian. "The Kill Chain: Defending America in the Future of High-Tech Warfare". Hachette UK, 2020. pg. 590.
64 Freedberg Sydney J., "Kill Chain in the Sky With Data: Army's Project Convergence", Breaking Defense., September 14, 2020, https://breakingdefense.com/2020/09/kill-chain-in-the-sky-with-data-armys-project-convergence/
65 Cited in Eugene Gholz, Benjamin Friedman & Enea Gjoza (2019) Defensive defense: A better way to protect US allies in Asia, The Washington Quarterly, 42:4, 171–189, DOI: 10.1080/0163660X.2019.1693103, 176.
66 Trevithick Joseph, "This Is What Ground Forces Look Like to An Electronic Warfare System and Why It's A Big Deal", The Drive. The Drive, May 11, 2020, https://www.thedrive.com/the-war-zone/33401/this-is-what-ground-forces-look-like-to-an-electronic-warfare-system-and-why-its-a-big-deal
67 Gartzke Erik & Jon R. Lindsay, eds. "Cross-domain Deterrence: Strategy in an Era of Complexity". Oxford University Press, 2019. pg. 77.
68 Ibid, pg. 8–9.
69 Ibid, pg. 133.
70 Cited in Ibid, pg. 134.
71 Carter Ashton B. (1986) Satellites and anti-satellites: The limits of the possible, International Security, 10:4, 46–98. Accessed November 13, 2020, DOI: 10.2307/2538950
72 Ibid.
73 Gartzke Erik & Jon R. Lindsay, eds. "Cross-domain Deterrence: Strategy in an Era of Complexity". Oxford University Press, 2019. pg. 135.
74 Patrick Tucker, "China, Russia Building Attack Satellites and Space Lasers: Pentagon Report", Defense One, February 12, 2019, https://www.defenseone.com/technology/2019/02/china-russia-building-attack-satellites-and-space-lasers-pentagon-report/154819/
75 Cited in Gartzke Erik & Jon R. Lindsay, eds. "Cross-domain Deterrence: Strategy in an Era of Complexity". Oxford University Press, 2019. pg. 135.
76 Ibid, pg. 137.
77 Brose Christian. "The Kill Chain: Defending America in the Future of High-Tech Warfare". Hachette UK, 2020. pg. 195–197.
78 Ibid, 464.

79 Kawasaki Charlie, "6 Ways AI Can Make Sense of Sensor Data in 2020", C4ISRNET, February 14, 2020, https://www.c4isrnet.com/thought-leadership/2020/02/14/6-ways-ai-can-make-sense-of-sensor-data-in-2020/
80 Alderman Ray, "C4ISR Systems and the Kill Web", Military Embedded Systems, December 20, 2018, https://militaryembedded.com/comms/communications/how-the-kill-web-manipulates-time-and-space
81 Gartzke Erik, Jon Lindsay & Michael Nacht, "Cross-Domain Deterrence: Strategy in an Era of Complexity" In International Studies Association Annual Meeting, Toronto, 2014. pg. 17.
82 Zak Kallenborn, "In a Robot War, Kill the Humans", Defense One, August 27, 2020, https://www.defenseone.com/ideas/2020/08/robot-war-kill-humans/168038/
83 Brose Christian. "The Kill Chain: Defending America in the Future of High-Tech Warfare". Hachette UK, 2020. pg. 469.
84 Ibid, pg. 532.
85 Watling Jack & Sidharth Kaushik, "The Democratization of Precision Strike in the Nagorno-Karabakh Conflict", Royal United Service Institute, October 22, 2020, https://rusi.org/commentary/democratisation-precision-strike-nagorno-karabakh-conflict
86 Ibid.
87 Ibid.
88 Ibid.
89 Davis Malcolm, "Cheap Drones versus Expensive Tanks: A Battlefield Game-Changer?", The Strategist, October 21, 2020, https://www.aspistrategist.org.au/cheap-drones-versus-expensive-tanks-a-battlefield-game-changer/
90 "LORA Precision Strike Tactical Missile, Israel", Army Technology, 2020, https://www.army-technology.com/projects/lora-precision-strike-tactical-missile/
91 Colley Christopher K., "What Iran's Attacks on American Bases Tell Us about China's Missile Program", The Japan Times, February 14, 2020, https://www.japantimes.co.jp/opinion/2020/02/14/commentary/world-commentary/irans-attacks-american-bases-tell-us-chinas-missile-program/
92 Watling Jack & Sidharth Kaushik, "The Democratization of Precision Strike in the Nagorno-Karabakh Conflict", Royal United Service Institute, October 22, 2020, https://rusi.org/commentary/democratisation-precision-strike-nagorno-karabakh-conflict
93 Ibid.
94 Eugene Gholz, Benjamin Friedman & Enea Gjoza (2019) Defensive defense: A better way to protect US allies in Asia, The Washington Quarterly, 42:4, 171–189, DOI: 10.1080/0163660X.2019.1693103
95 Cited in James S. Johnson (2018) China's vision of the future network-centric battlefield: Cyber, space and electromagnetic asymmetric challenges to the United States, Comparative Strategy, 37:5, 373–390. pg. 374.
96 Watling Jack, "The Key to Armenia's Tank Losses: The Sensors, Not the Shooters", RUSI Defence Systems, October 6, 2020, https://rusi.org/publication/rusi-defence-systems/key-armenia-tank-losses-sensors-not-shooters#.X3ys27k8_34.twitter
97 Brose Christian. "The Kill Chain: Defending America in the Future of High-Tech Warfare". Hachette UK, 2020. pg. 440.
98 James S. Johnson (2018) China's vision of the future network-centric battlefield: Cyber, space and electromagnetic asymmetric challenges to the United States, Comparative Strategy, 37:5, 373–390.
99 Ibid, pg. 374.
100 Brose Christian. "The Kill Chain: Defending America in the Future of High-Tech Warfare". Hachette UK, 2020. pg. 131–133.
101 Rahul Tripathi & Manu Pubby, "Covid-19 Delayed Indian Exercise, Chinese Moved into Key Positions", The Economic Times., June 3, 2020, https://economictimes

.indiatimes.com/news/defence/covid-19-delayed-indian-exercise-chinese-moved-into-key-positions/articleshow/76163629.cms?from=mdr
102 Ganguly Sumit & Frank O'Donnell, "China and Pakistan Have Been Taking Advantage of India's Intelligence Failures", Foreign Policy, August 27, 2020, https://foreignpolicy.com/2020/08/27/india-china-galwan-intelligence-failure/
103 Gartzke Erik & Jon R. Lindsay, eds. "Cross-domain Deterrence: Strategy in an Era of Complexity". Oxford University Press, 2019. pg. 160–162.
104 Ibid, pg. 160–162.
105 Brose Christian. "The Kill Chain: Defending America in the Future of High-Tech Warfare". Hachette UK, 2020. pg. 454.
106 For more on this, see Hersman Rebecca, Reja Younis, Bryce Farabaugh, Bethany Goldblum & Andrew Reddie, "Under the Nuclear Shadow: Situational Awareness Technology and Crisis Decision Making", Center for Strategic and International Studies, March 18, 2020, https://csis-website-prod.s3.amazonaws.com/s3fs-public/publication/200318_UnderNucearShadow_FullReport_WEB.pdf?VJm_nrx2bVVeByYH38yx8YkDvvr1QZVW
107 Ibid.
108 Gartzke Erik, Jon Lindsay & Michael Nacht, "Cross-Domain Deterrence: Strategy in an Era of Complexity", in International Studies Association Annual Meeting, Toronto, 2014. Pg. 4.
109 Thomas Timothy, "Russia's Electronic Warfare Force Blending Concepts with Capabilities", Mitre Center for Technology and National Security, September 9, 2020, https://www.mitre.org/sites/default/files/publications/pr-19-2714-russias-electronic-warfare-force-blending-concepts-with-capabilities.pdf. pg. 4.
110 Ibid, pg. 7.
111 Trevithick, Joseph, "American General Says 'Adversaries' Are Jamming AC-130 Gunships in Syria", The Drive, April 25, 2018, https://www.thedrive.com/the-war-zone/20404/american-general-says-adversaries-are-jamming-ac-130-gunships-in-syria
112 Borghard Erica D. & Shawn W. Lonergan (2019) Cyber operations as imperfect tools of escalation, Strategic Studies Quarterly, 13:3, 122–145. Pg. 126–128.
113 Hersman, Rebecca, Reja Younis, Bryce Farabaugh, Bethany Goldblum & Andrew Reddie, "Under the Nuclear Shadow: Situational Awareness Technology and Crisis Decision Making", Center for Strategic and International Studies, March 18, 2020, https://csis-website-prod.s3.amazonaws.com/s3fs-public/publication/200318_UnderNucearShadow_FullReport_WEB.pdf?VJm_nrx2bVVeByYH38yx8YkDvvr1QZVW. p52
114 Cited in Ibid, pg. 30.
115 Quoted in Ibid, pg. 30.
116 Caitlin Talmadge (2019) Emerging technology and intra-war escalation risks: Evidence from the Cold War, implications for today, Journal of Strategic Studies, 42:6, 864–887, DOI: 10.1080/01402390.2019.1631811, 866.
117 This debate is discussed in Caitlin Talmadge (2019) Emerging technology and intra-war escalation risks: Evidence from the Cold War, implications for today, Journal of Strategic Studies, 42:6, 864–887, DOI: 10.1080/01402390.2019.1631811
118 Brose Christian. "The Kill Chain: Defending America in the Future of High-Tech Warfare". Hachette UK, 2020. pg. 529–530.
119 Caitlin Talmadge (2019) Emerging technology and intra-war escalation risks: Evidence from the Cold War, implications for today, Journal of Strategic Studies, 42:6, 864–887, DOI: 10.1080/01402390.2019.1631811, pg. 866.
120 Hersman Rebecca, Reja Younis, Bryce Farabaugh, Bethany Goldblum & Andrew Reddie, "Under the Nuclear Shadow: Situational Awareness Technology and Crisis Decision Making", Center for Strategic and International Studies, March 18, 2020, https://csis-website-prod.s3.amazonaws.com/s3fs-public/publication/200318

_UnderNucearShadow_FullReport_WEB.pdf?VJm_nrx2bVVeByYH38yx8YkDvvr1QZVW. pg. 32

121 Ibid.

122 Caitlin Talmadge (2019) Emerging technology and intra-war escalation risks: Evidence from the Cold War, implications for today, Journal of Strategic Studies, 42:6, 864–887, DOI: 10.1080/01402390.2019.1631811

123 Cited in Ibid, pg. 880.

3

UNMANNED SYSTEMS IN CONVENTIONAL FORCES

The previous chapter dealt mostly with the network given its centrality to the battlespace. This "kill web"-centric model of operation places the network on a pedestal in the sense that the network holds the key rather than the platform. But the current revolution in unmanned technology and systems introduces some important changes in the landscape of conventional forces, as they bring to the fore the character of the platform at the last mile within the network such that the platform characteristics become as important in the overall context as the network. The larger trends unequivocally point to a favourable doctrinal disposition towards large-scale absorption of unmanned vehicles (UVs) and unmanned autonomous vehicles (UAVs) in forces across the world. This is also motivated by a larger trend in military defence expenditure across nations and particularly Western states where developmental pressures have forced defence budget cuts even as pension budgets have risen continuously over time.[1] To tide over such dependencies investments in unmanned systems have seen an exponential growth in the last few years. A good number of states have in fact used UVs to conduct strikes including the United States, Israel, the United Kingdom, Pakistan, Iraq, Nigeria, Iran, Turkey, Azerbaijan, Russia, and the United Arab Emirates.[2] The United States and Israel have extensively used UV strikes with MQ-1 Predator, MQ-9 Reaper, and Heron platforms against non-state actors. Similarly, Pakistan has used its indigenous unmanned system Burraq against militants in the North Waziristan tribal region.[3] Many states have invested in their own domestic programmes to procure UVs and UAVs. Russia, for instance, is developing its own Orion drone system which it has tested in Syria and is said to be in the same class as the Reaper[4] and a heavyweight system called the Hunter-B.[5] India has an indigenous programme for Rustom-I and Rustom-II MALE UAVs.[6] China has the Caihong family of drones that include CH-3, CH-4, and CH-5, Wing Loong II, and AV500W.[7] A lot of other countries are adopting unmanned systems at a

DOI: 10.4324/9781003365976-4

feverish pace in their force structures with at least three countries – the United States, China, and Israel – exporting these systems and some, including India, Russia, and Pakistan, developing their own indigenous versions.[8] These drone systems range from low-endurance, low-altitude to medium-altitude and long-endurance to high-altitude and long-endurance categories catering to various sizes of unit echelons and mission requirements and span from cheap units that can be manufactured in numbers to high-end expensive units that parallel manned units in their expenditure. In most countries aerial drone programmes have been front runners, but states are also increasingly looking to using unmanned systems in the maritime and undersea domains.[9,10] Further efforts have also been underway in some countries to adopt current assets for unmanned warfare. The United States, for instance, has used "autonomy kits" to convert some of its former fleet of manned F-16s into UCAVs, 4–5 of whom can be controlled by an F-35 as a loyal wingman to the manned fighter.[11] While it is also developing two-seater variants of the F-22 Raptor where the co-pilot can control a loyal unmanned aerial vehicle that can engage adversary fighters or "strike guarded targets".[12]

The overall evolution of unmanned systems has left analysts with the answer for "when" but not of "what" or "how" on the impact of UVs and UAVs on the force structure of conventional forces. This debate has particularly been stoked on the utility and feasibility of some of the traditional platforms on the future battlespace that were identified with conventional military forces. But the debate also extends to the shaping of the doctrine and operational concepts motivated by their use. On one hand the unmanned revolution provides for dispensable numbers in combat that inflict and survive the high level of attrition, yet on the other hand the nature of the conflict and the winning condition of an instrumentalized conventional strategy engagement demand a uniquely tailored level of force, matching of weapons to targets with an overall objective of establishing escalation control within a finite duration of the engagement. There is tension between these aspects of technology and objectives that contribute to the problem of managing the trade-off between effectiveness in the battlespace and escalation and crisis progression. They also introduce the problem of balancing autonomy with human decision-making in command and control even as they are perceived to provide solutions to newer challenges that arise from long-range missile systems and operational concepts designed around them. This chapter will delve into the impact of UVs and UAVs that are a defining feature of the third nuclear age, the challenges and benefits associated with them, the demands they may place on the current force structures, operational concepts, command and control, and the intended and unintended impacts they may have on current conventional military strategies.

Integration streams

Two streams of thought exist in integrating unmanned assets into the force structure. The first is the "mothership" concept[13] that lets unmanned systems be at the forefront in operations while the mothership is a manned vessel that operates in the rear carrying multiple sub-drone units catering to specific operational requirements that can be controlled by a human or it can operate autonomously. They increase the mothership's range, protection, and effectiveness. Scenarios include patrolling of a wide area by a ship with multiple UVs (or UAVs). These UVs are controlled by dedicated staff that act as natural force multipliers for the manned mothership. There are inherent advantages in this method, as many current manned platforms can be adapted to this role rather than investing money into a new type of platform built from the scratch for this role.[14] As drones become smaller in size, they can accommodate more to increase the strike power of the mothership.[15] This allows the current force structure of military forces to adopt drones in ways that do not interfere with their current mission particularly after having invested considerably into them. This kind of integration of unmanned assets demands centralized control[16] and therefore even for largely dispersed assets they are not likely to be very autonomous in their conduct.

A second philosophy of integration comprises swarms where there is no centralized control; rather, swarms are composed of individually highly mobile and autonomous units that act in unison against a target more effectively and lethally such that the sum of the parts is greater than the whole.[17] Swarms are governed by simple rules that are followed by each individual unit leading to better cohesion and collaboration among all units in the swarm, and essentially without any form of external control, though this is still an evolving area of technology.[18] They overwhelm defences by increasing the scale of threat in all directions, both spatially and temporally in that they can attack and re-attack once an opening is found.[19] There is perhaps no limit to the strength of the swarm, and theoretically anything from thousands to zillions can be afforded in a swarm.[20] Above all swarming has the advantage of de-loading the command centre which can focus its attention, energy, and bandwidth on other priority tasks. A third way could be a blending of these two with a hybrid form of control and balance between autonomy of the drone units and human in the loop.[21] A likely balance suited for manned and unmanned teaming.

Traditional manned platforms

The low cost of researching, developing, and making expendable unmanned systems is likely to introduce a new threat into the battlespace that could test the feasibility and utility of expensive large platforms that can only be procured in limited numbers and cannot suffer attrition. This is applicable for all physical domains – land, air, sea, and undersea – as UVs and UAVs add to the threat environment as well as offer new capabilities in each of those domains. These impose new trade-off equations for some of the traditional platforms like tanks and aircraft carriers. These debates are discussed below, where the focus is on the vulnerabilities of these platforms to unmanned assets and what that may imply for the rest of the force potentially demanding new approaches to tactics, capabilities, and their overall contribution to the force structure.

Land

On the ground, the debate about the utility of the tank in the battlespace has gained strength due to their apparent vulnerability against UAVs and potential swarms. The tank losses suffered by Azerbaijan in the Nagorno-Karabakh conflict in particular have warmed up this debate.[22] In this conflict Azerbaijan claimed to have destroyed 130 tanks while Armenia claimed it had destroyed at least 130 tanks.[23] Neutral observers had put Armenian losses at 185 T-72 tanks and Azerbaijan's at 22 tanks, with much of the losses being inflicted by kamikaze drones that self-destruct after flying to the target.[24] Analysts have suggested other arguments that potentially account for these tank losses on both sides other than the UAVs dominating the battlespace. First, many have suggested that much of these losses were in fact inflicted from anti-tank landmines and other anti-tank guided munitions.[25] Second, it has been suggested that the potential lack of training, particularly on the Armenian side, was responsible for the higher tank losses rather than the actual vulnerability of the platform to UVs. As witnessed in many videos the formations in which tanks were deployed made them easy targets; this included moving in tight clusters that prevented them from moving efficiently and lack of dispersion making strikes more effective against them.[26] Along with this the terrain in which these tanks were operating was supposedly a difficult terrain that needed sufficient expertise[27] that both sides lacked. Finally, the most crucial factor was perhaps the lack of necessary air cover and capability for counter air attacks on both sides.[28] In addition, the clever use of 11 slow Soviet-era An-2 aircraft by Azerbaijan was exemplary. Azerbaijan had modified these aircrafts to be used as drones within Armenian territory to trick their limited air-defences to fire and reveal their positions.[29] These were then taken out rather easily by removing the air cover for all ground forces, including tanks against UVs.

Analysts have therefore suggested that the tank's relevance has certainly not ended as they retain their utility in the age of air power and are likely to do so in states that have invested considerably in joint forces, doctrines that emphasize aerial protection for land units with networked and layered air defences along with capable air forces. A massed tank formation that can be protected against aerial threats can still provide the "assassin's mace" on the land frontier. Tanks can therefore be of significant utility in the escalation control and intra-war deterrence stage of the conflict on land. The answer is not likely to boil down to spending on one tank versus investing the same money for numerous cheap UVs that could potentially present the same level of threat to ground troops. This is because tanks will continue to retain their unique advantages in the land domain, and they might see more investments in counter-drone capability in the form of laser or other soft-kill methods like electronic capabilities besides operating under aerial protection. It is likely that the tank unit size might dwindle, or forces might go ahead with the option of "optionally manned" tanks, but this will necessarily be tailored to adversary force structure, nature of terrain, and the theatre of conflict. US Marine Corps, for instance, decided to phase out tanks in favour of wheeled High-Mobility Artillery Rocket System launchers and more drone squadrons to prepare for island warfare in the Western Pacific.[30] But the same may not be true in the case of forces that share land borders not excluding hilly terrain as seen in the India-China crisis in 2020 where both sides engaged in large-scale deployment of tanks.[31] Tanks could be employed with specialized anti-drone units, or this capability could be embedded at the level of the platform itself. But in a kill web with high situational awareness where concealment is impossible amidst thermal or infrared cameras and battlefield radars tracking by UVs may be unavoidable, integration of short-range air defences and electronic warfare units may therefore be needed as a standard capability for ground units including tanks.[32] Dispersion could be practised before getting in direct contact with the adversary where traditional capabilities and role by the land units can be realized.[33] The challenge is likely to be at the point of transition when dispersed units start concentrating.[34] Therefore UV detection and taking out adversary UVs, integration of electronic warfare capabilities will have to be a necessary part of the overall capability.[35] Small optimized ground force and tank units well defended from air are more likely to survive the threat of distributed unmanned aerial vehicles.[36]

Sea and undersea

In the maritime domain the force structure debate has revolved around utility of large aircraft carriers amidst the threat of hypersonic carrier killing missiles, drone swarms, and submarine threats.[37] Aircraft carriers are a major capability addition to the surface fleet as they carry aerial power and are essentially mobile. They can help project air power far away from the shore, and unlike the land-based aircraft bases, they can be closer to the area of interest. This mobility of the carrier allows fighter jets to maintain a better tempo of operations at the theatre of conflict in the maritime arena, meaning they can spend more time at the mission.[38] Land-based aircraft bases do not maintain the same tempo far away from land because those aircraft need to have enough fuel to return to the base safely. Therefore, time spent at the target is considerably reduced.

The most significant threat to the carrier in future may come from anti-ship missiles.[39] Missile systems like DF-26 IRBM with their 4000 km range and high precision have been touted as "carrier killers".[40] However, targeting mobile carriers in the vast expanse of the seas is not easy as tracking the carrier itself is problematic despite its huge size and radar signature.[41] At long ranges these missile systems also require continuous updates to effectively hit the target.[42] While satellites can track the carriers to have targeting quality information these satellites must operate in the Low Earth Orbit (LEO).[43] This would need a big constellation of satellites and an earth-based system for maintaining continuous track which could be fairly difficult. Another option could be to use unmanned or manned systems to track carriers; however, these would be at close range of the carriers that do operate with a multilayered defence that includes both hard kill and soft kill options, thus making these airborne systems vulnerable.[44] Finally, carriers are also highly survivable and capable of operating even after taking a hit, though overall the carrier might be forced to operate from greater distance against such threats. It is also likely that carriers might increasingly adopt UVs and UAVs in their air wing instead of only a manned fighter wing to fly combat air patrols.[45] The rest of the surface fleet is also likely to see a greater integration of unmanned systems, with countries investing in unmanned surface ships.[46] The real potential transformation for aircraft carrier could likely see a new role for operations after the first or second wave of a future attritional conflict with a peer competitor has exhausted coastal defences, aerial and land-based sensors, and missile batteries, where it could then, similar to a tank in the land domain, deliver the critical blow with its manned and unmanned assets.

Finally, we have the sub-surface where the ramifications of the use of unmanned platforms are likely to straddle the divide between conventional and nuclear realms more than the other domains. The United States, for instance, is developing the Sea Hunter unmanned surface vehicle (USV) and "Orca" large unmanned underwater vehicle (XLUUV).[47] India and China[48] have also gone ahead with their own programmes for these vehicles intended to perform reconnaissance, mine placement, and kamikaze attacks. Deep-sea drones that can sport active sonar and non-acoustic detection methods could make it hard for traditional submarines to hide.[49] However the current unmanned undersea platforms lack combat capability against the traditional lot of nuclear attack submarines and conventional submarines that are faster, quieter with higher endurance and are unlikely to be easy prey.[50] Yet these unmanned undersea platforms do make manned platform operations in hostile waters a very dangerous and risky proposition. In fact, many of these manned platforms might see investments at the platform level to play the role of manned motherships that can deploy many more swarms of drones.[51]

By some estimates the synergy between the unmanned and manned surface and undersea sensors could match the sensor networks of air-land and therefore mirror the kill web on land and in air, in the sea.[52] Though controlling a large number of undersea unmanned systems with their slow data transmission rates is not practical, they could be used in conjunction with other means like sonar networks and satellite-based detection systems, among others.[53] The resulting maritime domain awareness could impact the sub-surface component of the strategic forces heavily given this is one leg on which nuclear deterrence and second-strike capability have traditionally hinged. However, this is not likely to be a problem only for the defendant as the aggressor may also need to ensure that unmanned systems retrieved or recovered by hostile forces do not mistakenly give away the location of their own mothership manned vessel which could then be vulnerable to a strike.[54]

The overall impact on naval force structure is likely to lead towards distributed lethality. The US Navy, for instance, wants to have proportionately less large surface combatants (cruisers and destroyers) and more small surface combatant ships (frigates and littoral combat ships) with a significant number of large, unmanned vehicles.[55] There are plans to install capability to launch swarm aerial drones from sub-surface and surface-based unmanned ships.[56] This is a new concept of operation known as distributed fleet operations where navies across the world will shift from depending on a few large ships for their aggregate capability towards combining a large number UAVs and UVs that increase the adversary's targeting challenge with a larger number of units to detect, identify and track, and reduce the loss in aggregate capability from the loss of one individual ship.[57] For the undersea component of forces, however, only UAVs and not UVs are likely to be the choice. This is due to the difficulty of undersea communications as radio waves do not travel far underwater, therefore requiring that unmanned undersea vehicles are autonomous in operations and therefore needing progress in artificial intelligence as a prerequisite.[58]

Air

In the air domain the force structure debate focuses on whether the manned-unmanned teaming can eventually give way to manned fighters being completely substituted by unmanned systems. Experiments in this area have yielded some interesting results which, however, need to be read keeping in view the context. A virtually simulated dogfight, for instance, using guns between a human piloting an F-16 and an artificial intelligence–operated F-16 led to the Falcon artificial intelligence managing to defeat the human pilot 5-0. The fights were set up for different Basic Fighter Manoeuvres (BMS) within visual range (WVR) at different altitudes. The artificial intelligence consistently disregarded the flight safety rules in this virtual battle by getting within 500 feet of the human pilot during the pass. This 500-foot separation is deemed necessary to avoid debris from the plane being shot at as well as to avoid mid-air collisions.[59] The artificial intelligence also took some sharp mid-air manoeuvres that were not possible for a human pilot in a real fight. In a real fight g-force can be a limiting factor for a human pilot which was not the case with the artificial intelligence.[60]

The artificial intelligence did consistently demonstrate a faster OODA loop in decision-making than the human pilot by attempting to take a face-shot while the human pilot did not attempt to do that. This artificial intelligence agent was fielded by the Heron systems that employed a reinforcement learning technique during its training that assigns weight to each action and updates it after every iteration of the simulation it learns by trial and error.[61] It eventually went through some four billion simulations which translated into "12 years of experience"[62] compared to the human pilot who had about 2,000 hours of flight experience.[63]

The artificial intelligence pilot was not allowed to learn from its real-time experience during the virtual dogfight, and it operated under the perfect information about the scenario and aircraft parameters.[64] It is therefore possible that had it faced a new strategy, such as a two-on-one fight or lack of information about the aircraft or suffered a degraded situational awareness due to an electronic attack, the outcome could have been different. This was evinced in the fifth round when the human pilot changed his tactic and was able to survive much longer in the fight as compared to the previous rounds.[65] Yet the simulations results did show the potential of artificial intelligence–enabled unmanned expendable drones, loyal wingman and manned fighter plane controlling multiple drones.[66] Though perhaps there is some time before unmanned systems could mirror a human pilot in its overall capability and skill, there is very little doubt about the trajectory of the future force structure here given the potential capability and applications. Particularly swarming and kamikaze drones are likely to be a very important part of future air doctrines and capabilities. These are also likely to incentivize countries to work on the loyal wingman concept where these are expendable unmanned systems that mirror the capability and performance of the manned fighter.[67,68] Loyal wingmen can penetrate adversary air bubbles to deliver munition on the target while manned fighters can stay at standoff range[69] enabling a host of new operations and missions.

Trade-offs

The crux behind the motivation to integrate unmanned systems in the force stems from the fact that they place the human operator at greater distance from harm while these systems themselves are expendable in situations. The direct benefit is that at the frontlines of the conflict, machines substitute for humans making use of force more tenable politically and financially. In absolute terms given human involvement particularly in the case of UVs the on-ground operator teams and repair or maintenance groups that keep these systems running does not necessarily diminish the role of the humans in the overall structure but the exposure of humans to the frontlines of the conflict is definitely diminished. Supervisory roles for humans are most likely to stay as information is sent to them to act upon.[70] In fact the man-to-machine ratio is likely to be higher as even in simulations it has been observed that two-person system operators have shown better efficiency than single person–operated unmanned systems.[71] This further depends on the number of payloads carried by the unmanned system and the assigned roles given the limited cognitive capacity to operate them efficiently and simultaneously.[72]

So, at a fundamental level though they decrease the exposure of troops to the frontlines, this is not necessarily a case of saving expenditure or substituting the role of human operators who may still be required. In fact, cost comparisons between unmanned and manned systems are not straightforward. Many UAVs, for example, have ground operating teams which incur a regular expenditure similar to the manned platforms. Some systems could even have a greater cost simply due to greater capability; for instance certain UAVs could have greater cost due to carrying more sensors than the manned counterparts or when the manned counterpart has a greater operational cost due to higher fuel consumption; as a result, they are bestowed with greater speed helping them manage air-to-air combat unlike a UAV.[73] Cost comparison therefore as a basis to extrapolate future force structure changes may not fructify well.

That said, there are inherent advantages unmanned systems bring to the fore. The endurance that unmanned systems bring to the operations is, in most cases, limited only by the fuel burn rate. The human teams operating the system can afford to work in shifts while platform can conduct persistent operations. In addition, because there are no humans to carry on the platform or in some cases very minimal level of manned troops are required to operate these platforms on board, they offer more real estate to pack in more sensors, on-board computing power, and communication equipment, etc.

However, some natural and man-made developments might still impede their integration in the force. First, the lethality of a single unmanned system against a hardened military target is not very potent given their limited payloads relative to manned fighters.[74] A second key barrier is electromagnetic interference that often impedes robots from working properly in the commercial space, as the steering mechanism is often dependent on GPS signals.[75] Many such interferences are capable of jamming, and specialized devices that can do that in small form factor are now available very cheaply.[76] In fact, as countries have researched, many have developed specialized radiofrequency bombs and guns that can create massive electromagnetic pulses that they have now harnessed the capability of even frying the circuits in these systems.[77] Though UAVs have the advantage of loitering time over manned counterparts, they are also more vulnerable to electronic hacking[78] and sophisticated layered defences.[79] In 2011, MQ-1 and MQ-9 drones in the Middle East region were infected with hard-to-remove malware, exposing the vulnerability of unmanned systems to possible hacking and being forced to go rogue on their own teammates.[80] Many military communication channels are dependent on commercial components or go through commercial networks, and these are not developed in a way where they are resilient against such network breaches.[81] Ideal segregation between commercial and military networks is desired but not always achieved in many cases. Another problem relates to higher bandwidth consumption where images or other media from the unmanned system need to be transferred to a command post but where it must compete with audio input from other 20–25 command posts.[82]

Overall, as unmanned systems grow, investments in myriad form of counter-drone technology for point as well as area defence are also growing. Development of counter-drone capabilities at the level of the platform and as part of a combined arms unit could particularly see rapid growth in the coming years. There are, for example, plans to outfit surface fleets with laser systems to combat aerial UVs and UAVs.[83] Other than that, states have also invested in short-ranged kinetic hit systems for the protection of tactical units from drones. However, the most effective counter for swarms currently seems to be microwave weapons that could fry electronic circuits and are small enough to be mounted on trucks to defend bases, command posts, and logistics depots.[84] Unlike lasers that act like a "long-range blowtorch", microwave radiation only scrambles the electric circuitry.[85] These are more useful against large swarms because they deliver the effect along an arc unlike lasers that act along a narrow thin beam.[86] The trade-off is that microwave weapons are shorter ranged than lasers. This is likely to ensure that these capabilities are used in conjunction to protect against the full spectrum of unmanned systems threat. These counter-drone systems are not without weakness. Lasers can take out only one drone at a time, or they would require a lot more equipment for multiple beams which could become cost-prohibitive.[87] These laser beams would also have to be at the target for several seconds to inflict major damage.[88] Likewise electromagnetic weapons might be useful against unmanned systems, but they come with the trade-off of spiking up the electromagnetic signature of the unit giving away the target to the adversary. Second, mobile land units don't often have the generator capacity to take out such drones when they operate in a swarm formation. For instance, Russian units have come to embed electronic warfare and area denial capabilities at the level of the brigades and division, but such capabilities, although enough for squad sized drones, may not be enough to tackle a swarm. A much greater problem is the inability to scale up these defensive technologies for a large-scale unmanned attack, so offence dominance increases as investments scale up in both counter-drone systems and drones especially at the tactical level.

The operational advantages for unmanned systems materialize in terms of lethality and damage infliction capability. Particularly loitering kamikaze drones, for instance, fill a role that falls somewhere between missiles and other armed drones.[89] These loitering munitions sometimes also have man in the loop feature because they are equipped with cameras that can relay video right until the point of hitting and therefore, they can be maneuverer until even the terminal phase of the flight.[90] UVs and UAVs are therefore likely to become organic across the force[91] as they saturate the battlespace with sensors, shooters, and sensor-shooter packages.[92] Among other factors UVs could break the quest for the qualitative edge that had consistently led to long acquisition cycles and high development and sustainment costs.[93] UVs and UAVs are likely to replace long gestation periods with short cycles of development for large numbers of inexpensive unmanned systems instead of small numbers of very expensive, but vulnerable, manned platforms.[94] Unmanned systems will therefore reintroduce mass into the battlespace that had been in decline since World War II, after which capital-intensive manned platforms become the norm and the high per unit cost discouraged large force structures.[95] Further, regardless of the domain, they offer inherent advantages like endurance for persistent operations that manned platforms may not be able to emulate. This is a mixed picture with a race between the shield and the sword, but their heavy integration across domains seems a foregone conclusion that will shape both the future force structure and design.

Unmanned systems and ICS

Unmanned systems do place the operator away from harm on the frontlines, but that may not comport well with an instrumentalized conventional strategy where the objective is to shape behaviour and the operation environment in one's favour gradually. For, one, unmanned systems do not increase the stake for an adversary who might be more willing to take lethal action against such targets in the hope that eliminating unmanned forces carrying out a territorial or behavioural fait accompli action is not the same as firing at human troops carrying the same action and therefore might help create a deterrent or preventive precedent. Own perception of red lines and lower-level thresholds with unmanned systems may not be perceived in the same vein by the adversary. Repeated use of purely unmanned systems could in fact create new thresholds for shooting down "intruders", manned or unmanned, over time. So, by their very nature of being expendable they do not support peacetime coercion as is often experienced in a successful instrumentalized strategy with manned troops. For instance, as US-Iran tensions grew on 20 June 2019 Iran shot down a US RQ-4A Global Hawk spy drone that had intruded into Iranian airspace.[96] The United States described the incident as an "unprovoked attack". A few months later in January 2020 the United States conducted a drone strike, killing Iranian military commander General Qasem Suleimani, the head of Iran's Quds Force.[97] Both incidents marked a major escalation and breaching of earlier established thresholds between the two conflicted states. One incident had a drone target, whereas the other perpetuated itself in the form of a drone strike. This was later exacerbated by the Iranian reaction to the drone strike in the form of a missile attack on US military bases and subsequently, but not surprisingly, the inadvertent Iranian shootdown of a civilian Ukraine International Airlines flight PS752 described as an "human error".[98] The crux here is that repeated use of unmanned assets in defensive and even in offensive roles can lead to an inadvertent dilution of established thresholds as the gravity associated with taking lethal action against unmanned units is not the same as with manned units. Subsequent escalation could then get compounded by a greater desire to resort to missiles, manned air strikes, or further drone strikes. In that sense they tend to increase the pressure to vertically escalate in a conflict.

That said, a combination of manned and unmanned forces can reduce the potential pitfalls and avoid lowering the threshold for conflict while combining their individual strengths. Such a force could sustain and defend a fait accompli better than a purely manned force. It can occupy favourable military positions far more swiftly with a reduced logistics train. Unmanned systems can deliver crucial war material to frontline units. This utility of the unmanned units is likely to be contingent upon both technology and the terrain. Higher-altitude regions with extreme windy conditions might not enable use of unmanned systems for logistics support. Unmanned forces are likely to have a very important role in the escalation control stage of ICS once manned troops are positioned to move into escalation control or intra-war deterrence stage. In this stage the expendable nature of unmanned systems and their ability to inflict damage could aid status quoist forces and deter forces seeking revision of the established fait accompli. Such forces need not exhibit a mixed force structure utilizing manned and unmanned assets; instead, such forces could afford to be purely unmanned in structure with human relegated to operating them remotely when needed and focusing more on imagery and intelligence interpretation or force application.[99] Positioning unmanned systems in large numbers in this stage could convey higher willingness to resort to use of lethal force. If these systems are operating in a kill web, the threat will be even more pronounced. The downside of the use of unmanned assets is that their expendability could incentivize kinetic action and decrease crisis-stability while enabling speedy progression to larger-scale conventional operations, prolonged engagement, or the conventional-nuclear transition point. This contradiction between the efficacy of unmanned systems versus the necessary sequential, controlled and gradual development of ICS is a thin balance that nuclear weapons states may have to consistently figure out in crisis situations.

Lower echelons, mission command, and dispersion

Command and control of the unmanned systems in the kill web is likely to present opportunities, challenges as well as greater risks. Fundamentally command and control encounters three new introductions that are likely to characterize the third nuclear age. The first is the challenge posed by the instrumentalized conventional strategy as the resultant peacetime coercion enforces a change in mission objectives. The second is the challenge posed by the kill web and the cross-domain nature of threats on operations. And, finally, unmanned systems and unmanned autonomous systems bring their own unique attributes into the threat mix that add the challenges to military command and control. One key takeaway from the previous text was that the tactical and operational levels of the conflict become much less distinguishable and in fact get enmeshed in ICS. In an instrumentalized conventional strategy, the higher echelons of command and control have to remain more engaged at the tactical level, as operational-level consequences are inflicted over time and in a gradual manner. In addition, kill web is likely to push combined arms capabilities at the tactical level, and as a ramification a lot of the combat power is likely to be furnished by joint assets at a much higher level of command,[100] with a majority being unmanned and autonomous assets.

This places a requirement on the command and control architecture to be more agile and responsive in supporting tactical echelons and to have the capability to introduce operational-level deterrence to conflicts at tactical levels. These lower echelons that are going to become more important could be at the brigade- or even company-level formations. A conflict that engages company formations is likely to have repercussions for brigade-level echelons, and those at brigade levels are more likely to pull in division-level interventions. This implies that as theatres become important, theatres of war, or theatres of operations or even joint operation areas, are likely to become the focus and define the scale of the conflict. Second, shaping operations are likely to be of more consequence and need to be accorded higher priority. It is very likely that decisive operations where kinetic engagement is required may in fact never take place because main effort will be directed towards shaping and counter shaping the operational environments yielding tactical and in time operational and strategic consequences. The role of cyberspace, space, or other kinds of cyberspace operations becomes pertinent for higher echelons as the effects generated at the tactical level can have an impact outside the area of responsibility for tactical units.[101] The fact that a variety of unmanned systems are being designed and developed for the lowest level echelon of the forces[102] is therefore not just capability addition at these echelons but a natural consequence of the fact that lower echelons in the force have become important to stability and the deterrence equation between nuclear weapon states as a consequence of instrumentalized conventional strategy.

Concepts like anti-access/area denial (A2AD) that involve the use of long-range ballistic, hypersonic, and cruise missiles along with early warning and long-range radars, integrated air and missile defence systems, long-range reconnaissance satellites and aircraft[103] fall in this bracket of shaping operations that manipulate the operational environment over time by enabling tactical moves. Once the operational environment is set instrumentalized, conventional strategy is enacted and the outcome is forced on the adversary. All of this transpires within a cross-domain kill web composed of manned and unmanned assets where adversary forces in the deep area are not out of contact and these set up the environment for the second stage of intra-war deterrence in ICS. The capabilities afforded in terms of situational awareness, data-centric operational ecosystem with sufficient precision, range along with integrating unmanned assets all feeding into a cross-domain system of sensors and shooters within a web severely restrict the space for very large-scale offensive missions and manoeuvres. This is likely to force militaries to "isolate parts of such system to enable tactical manoeuvre at operational depths".[104] Corps-level echelons therefore remain engaged in both physical and temporal operations in the deep area to shape the operational environment for frontline forces[105] that inflict the eventual fait accompli. The temporal dimension extends to predicting future courses of adversary action acting accordingly to prevent them while the physical dimension includes acting against those forces which may not be at the frontline but are present in the rear and could inflict damage in the future. As mentioned in the previous chapter, kill web materializes in terms of greater situational awareness, exchange, and integration of information among different domains in a data-centric architecture. To this "more visible" or "less opaque" battlefield it adds precision and range as the last-mile threats are embedded in the form of various long-ranged and short-ranged missile and unmanned weapon systems. This growing range and precision of surface- and air-launched weapons along with kill web's cross-domain sensing and shooting ability is likely to reinforce an ongoing trend in ground forces forcing them to operate with greater, rather than less, dispersion.[106] This will place greater pressure on coordination and manoeuvre challenges in a time-sensitive environment.[107] Operationalization of the kill web will therefore demand a command and control architecture that can cater to widely dispersed manned and unmanned forces or doctrines that do not depend on manoeuvre.

The demands of kill web are, however, likely to pressure the traditional doctrine of mission command,[108] the nodal point that sits at the heart of executing dispersed operations. These multi- or cross-domain operations demand a command and control arrangement that is responsive to cross-domain operational needs, convergence of fires, and high operational tempo.[109] Mission command concept has traditionally allowed ground commanders to work independently within the broad guidance of the higher echelons. Dispersal of own forces has the motive to strive to create multiple problems for the adversary with dispersed, but networked, units comprising both unmanned and manned assets. As it strives to overwhelm the adversary command and control and decision-making psychologically, with dispersion of its own forces such deliberate tactics also place pressure on own command and control to maintain cohesion and coherence between different units to the point that dispersion may prove counterproductive.[110] Mission command that enables initiative at the lower echelons also inevitably places a greater pressure on convergence and synchronization requirements that become prerequisites due to operationalization of the kill web and cross-domain functionality at the lower echelons.[111] The risk of this incoherence becomes greater with the kind of delegation that is awarded in the mission command concept. It is important, however, to note that greater dispersion of forces is a factor that is inevitably tied to the scale of the battle and whether instrumentalized conventional strategy is operative, shaping the nature of confrontation, among other factors, like scale, iterations, and trajectory of crisis progression.

Unmanned systems come to the rescue here as they reduce the need for manoeuvre.[112] In fact countering as well as advancing an instrumentalized conventional strategy will require absorbing and making efficient use of unmanned vehicles and unmanned autonomous vehicles in the most efficient manner as part of manned-unmanned teams, swarms, or unmanned systems alone. Unmanned systems, particularly airborne drones, for instance, become important as they not only offer unmatched endurance with a long dwell time but also can create a backlog of events which could prove useful in monitoring shaping operations that seek to change the operational environment over time.[113] An environment saturated with numerous unmanned systems in conjunction with kill web could potentially create 24×7 visible frontlines and battlespace conditions, where every and any movement of troops can be detected and acted upon in a time-bound manner contingent upon military organization having the culture and the structure for responding to the alerts generated by the web. However, the use of UVs and UAVs will also impose new requirements on the traditional conventional command and control systems that must imbibe new rules of operational conduct discussed in the next section.

Joint all-domain command and control (JADC2) and autonomy

Conventional forces dispersed across multiple domains in a kill web will require a unified digital command and control approach. The genesis of the US concept of JADC2 lies in controlling this dispersed force that operates in a unified single entity, the battlespace. JADC2 is to span across echelons and place requirements of cloud computing at the operational level and at the edge or forward positions for control of forces in all domains. The US Army, for instance, has tested a chain of artificial intelligence, software, and UAVs to take sensor data from multiple domains and used them to convert it into targeting data along with selection of the best weapon system to respond to the threat within seconds.[114] China too has been working on linking its army and air force units into a unified command mechanism to connect air defence radars, PLA ground units, and communication.[115] It is imperative for JADC2 to communicate within fractions of a second, to process sensor intelligence and data, evaluate that data, and decide which systems have priority and control over the target in fractions of a second. This task has to be assigned to the network which then adjudicates which platform and which domain will deliver the effect, whether kinetic or non-kinetic, depending on reaction times and attack objective. Or it could be performed by a human which is advised by artificial intelligence, where choices are already filtered by the autonomous kill web. Either way autonomous decisions must sit at the heart of JADC2 to ensure that the required effect can be delivered in an efficient and timely manner. Intelligence and algorithms will identify targets for manned command centres and recommend decisions and courses of action. In JADC2 these algorithms work to pair target sensors with the shooters who can then deliver the kinetic or non-kinetic effect, where both manned and unmanned platforms are plugged into the same kill web across domains.[116] This framework is more than about just connecting assets or greater situational awareness; it is the essence of how decisions are made and arrived at. This framework is likely to comprise operational-level C2 nodes that can utilize capabilities across domains rather than capabilities from one domain or only one service and therefore be more mission function capable than being domain-centric or service-oriented in terms of effects in the battlespace.[117] This requires dynamic update of data across the kill web for units, their capabilities, threats, and the weapon or platform performance potential for acting against that threat.[118] Edge processing might allow this to be completed at the level of tactical units where these results are then transmitted to operational-level C2 nodes.

This autonomy embedded in JADC2 will enable it to decide which units are available and which ones are best positioned to complete a task in theatre-wide conflict.[119] In addition at the level of the platform introduction of certain kind of unmanned systems, particularly in the undersea domain, is likely to place a greater emphasis on autonomy than remote operated systems with "human in the loop" because radio waves can't travel far under water and large-scale control of unmanned units or swarming cannot effectively be directed by a human entity.[120] This presence of artificial intelligence on both the platform, network, and perhaps in the future on the weapon is likely to ensure that the strike can proceed without human intervention.[121] This ability to process data on the move from various platforms in multiple domains could translate into fundamental advantages of speedy decision-making.[122] Autonomy-enabled integrated battle management could then, for instance, be the key to finding and targeting missile batteries helping counter A2/AD doctrines.[123] It could also act on some pre-specified courses of action for certain contingencies, these may include "detection of an air-defence threat to a critical sensor with the option of engaging it with counterfire or moving out of that airspace".[124] Autonomy on unmanned vehicles for instance could allow them to operate in A2/AD zones even when communication links or remote-piloting links do not survive allowing them to complete their mission by resorting to pre-programmed directives.[125] While automation in command and control may bring about better efficiency by reducing the intelligence disconnect between different echelons and allow for better contingency planning and therefore allow different units to harmonize their adjustments for re-tasking.[126] This could potentially extend to integrating logistics with operations with comprehensive logistical tracking and better information commonality.[127]

That said, there are potential downsides of networked automation. The use of unmanned assets that can provide commanders with live video footage also means that they often provide the incentive for commanders to intervene during tactical missions leading to micromanagement.[128] Peter explains that

> "the traditional concept of a military operation is a pyramid, with the strategic commander on top, the operational commanders next, and the tactical commanders on the bottom layer. With the new technologies, this structure isn't just being erased from above, with strategic and operational commanders now getting into the tactical commanders' business. It is also endangered from the sides. As one drone squadron officer explains, a major challenge in the command and control of reach back operations is their simultaneous location in multiple spaces."[129]

In the concept of JADC2, autonomy aids the human commander and then realigns and readjusts orders as the mission situation progresses or new intelligence emerges.[130] "Man in the loop" is considered essential and the application of artificial intelligence exists to aid the human commander. It is likely that even full autonomy could potentially only be used to aid human agency in decision-making as autonomy could present options and assist the human commander in diagnosing battlefield impediments and time-space implications for choosing the next course of action on the battlefield.[131] Greater autonomy in JADC2 therefore should not necessarily be seen as something that is at the cost of reducing the human agency in decision-making. But this is a carefully deliberated and experimented decision that is arrived for each state's unique threat spectrum. All nuclear weapon states are likely to evolve to some version of JADC2 to operationalize kill webs and control cross-domain application of force; however, there can't be any certainty that they will follow the same modus operandi of enabling the human commander and even if they do, what form and shape it might take given the unique operational- and strategic-level challenges each of them might be facing. Countries like China have talked about artificial intelligence helping in convergence of fires across domains but the extent to which artificial intelligence might aid or substitute human direction is unclear.[132]

The important aspect here is that unmanned forces as of yet are not stakeholders and only the human is responsible for the end-state.[133] Autonomy on the platform is likely to increase the challenge of doing so because the more autonomous a task becomes the harder it becomes for the human to intervene or react to a contingency because they do not maintain the necessary awareness for taking control of that process.[134] For instance an Israeli virtual battle management system is capable of supporting mission commanders but can also take over decision-making in contingencies where "incoming targets overwhelm human decision-maker".[135] This could be necessary in a time-sensitive or high-pressure environment where one may be dealing with, say, large swarms of unmanned systems that cannot be handled by a human. However, the risk that humans might lose the control over the tempo of the conflict and this might accelerate the progression of the crisis during an operation does exist.[136] This would be incompatible with the instrumentalized conventional strategy that often seeks controlled progression of stages and therefore human in the loop is likely to stay the preferred way of operation if instrumentalized conventional strategy is the preferred way of operating in the third nuclear age. Greater or lesser autonomy in the JADC2 will therefore be a trade-off between higher efficacy and control over crisis events and escalation. The scope for miscalculation will, however, exist in this space, if nuclear weapon states don't communicate or share best practices with each other.

Conventional-nuclear transition

As discussed earlier the greatest threat unmanned systems pose is that of lowering the threshold for conflict. In many ways use of unmanned assets may run contrary to what may be needed in ICS strategy that seeks gradual control over time and supports escalation control and intra-war deterrence in conventional contingencies. The very expendability that lends them efficiency and high performance in conventional combat also allows them to lower the threshold for miscalculation, false assumption of where the deterrence equation might stand, and trap actors into committed force postures that might take an accidental route to a bigger crisis. However there are other unintended impacts of unmanned systems and indirect implications that might impact the conventional-nuclear transition point in a conflict.

The first is the combination of unmanned systems, kill web, and JADC2 that offers exquisite ISR, platform, and networked capabilities makes it possible to target strategic assets like aircraft carriers, road or rail mobile missile systems, and nuclear weapons degrading second-strike capability and first-strike crisis stability.[137] This could have a greater impact on states with smaller nuclear arsenals, unlike the United State or Russia that possess large nuclear arsenals, making them nervous about their "assured" second-strike capability. The impact of unmanned systems directly on the nuclear deterrence equation between states could be most profound in the undersea domain as unmanned systems could render the undersea domain more transparent.[138] The impact is in terms of both nuclear and conventional systems. Many states that have conventional diesel-powered submarines have had an advantage over their nuclear-powered submarines in that relatively cheaper diesel submarines can be quieter and hide in the littoral waters and therefore threaten the more costly nuclear submarines as they are engaging them in the deep blue sea.[139] Nuclear submarines can't afford to use active sonars to find these submarines as they have the effect of giving away own location. Unmanned systems here can offset that advantage as they can afford to use active sonars to look for these adversary submarines without letting the threat reach the mothership.[140] This tactic could nullify nuclear capability of states with a nascent undersea or sea-based leg of the nuclear deterrent, particularly those who deploy nuclear weapons on short-legged conventional submarines in bastion defence. As these could then be tracked in peacetime and in a crisis situation taken out to prevent them from coming into action, and further encourage conventional or nuclear counterforce doctrines. A situation where all three legs of the nuclear triad are simultaneously vulnerable has not transpired in history. But the fact that such a situation could transpire in future, particularly where artificial intelligence prediction could increase the chances of being successfully tracked especially for smaller and weaker NWS and those that share land boundaries, could put them in a very distrustful position[141] even in conventional conflicts. Further, even in conventional operations artificial intelligence–augmented aerial swarms could possibly take out ground-based air defences, C3 systems, or early warning systems that could pressure a weaker state into "use or lose" dilemma.[142]

Many nations that have manned platforms for delivering nuclear weapons or in the nuclear strike role are likely to be more vulnerable in adversary kill webs where the aerial threat is posed by the fact that they can be detected early. Most pressing concerns come from the fact that autonomy-enabled swarms could in fact act as missile defence by defeating traditional nuclear delivery systems and potentially against hypersonic missiles.[143] Gigantic swarms could be used as aerial mines that could alter the trajectory of the long-ranged delivery systems or destroy them by colliding with them.[144] Further if states pursue unmanned systems for nuclear delivery roles combined with swarms to trick defences, this will pose platform ambiguity problems for the defender.[145] If there is no clarity on whether they should be seen as a conventional or nuclear attack platform, chances are they will degrade crisis-stability and force the defender state into drastic action.

The other set of risks originates in technological complexity of the current generation of platforms and networks. Particularly with unmanned autonomous vehicles, automation-assisted manned platforms, and over automated networks, complexity is also a factor in triggering crisis and miscalculations between nuclear weapon states. Platform-related complexity has in fact been witnessed in many forms in the era gone by. In one incident on 10 October 2007 during a military exercise an automated anti-aircraft gun MK5 having two 35 mm cannons began firing on its own after having appeared to be jammed.[146] It fired high-explosive shells at 550 rounds per minute revolving 360 degrees, during which its auto-loading magazines killed nine personnel and injured another fourteen seriously.[147] The problem was attributed to a "software glitch".[148] The possibility that such glitches grow with more artificial intelligence spanning complicated platforms, swarms, and UAVs could lead to unintended consequences is a possibility. Kill webs and JDAC2 are likely to add a new dimension of complexity to the already existing problem of having complexity at the level of platform. The possibility of false alerts or false warnings and the pressure to respond quickly, or the fact that the autonomy in the network could take over human control, leave a lot of room for losing control over events at a sensitive time or crisis situation. In 1960 an automated ballistic missile early warning system based in Greenland had "detected" a launch "with 99.99 per cent certainty" forcing NATO to go on alert. The alert was in fact caused because of the computer confusing a rising moon with a missile launch.[149] Singer notes that the possible ramification of this episode had it happened two years later during the Cuban crisis could have led to a nuclear exchange.[150] With artificial intelligence–augmented platforms and networked capabilities the risk that false detection, especially during a crisis, could lead to large-scale conventional war or a large-scale conventional war could transition to use theatre-based use of nuclear weapons and further into a strategic exchange is a possibility.

Conclusion

The overall contribution of unmanned systems in shaping the characteristics of the third nuclear age is immense. Even with the various trade-offs in current capabilities and their incompatibility with instrumentalized conventional military strategies, the perception around the advantage they could potentially offer in reducing the exposure of the human operator from direct threats and substituting their role at the frontline is seen revolutionary. Their contribution in enabling and operationalizing the kill web, newer concepts of operations like distributed lethality, and doctrines shape the essence of cross-domain sensing, tracking, and damage infliction. How the unmanned revolution is going to play out in terms of the traditional understanding around escalation, deterrence, first-strike stability, and second-strike capability, and even kill web, JDAC2 is difficult to predict at the moment given most concepts and capabilities are still evolving. It is, however, a given that unmanned systems will bring back mass into conventional conflict, forcing traditional platforms to adapt to the threat in capabilities at the level of the platform, force structure, and doctrines. It is very likely these platforms will take up new roles in a new doctrine or get resized rather than see termination altogether. The greater challenge for nuclear weapon states is likely to be catering to peacetime coercion using unmanned systems given they have the unique contribution of lowering the threshold for kinetic hostilities even as they provide more lethality and efficiency in targeting. The challenges for unmanned systems are also likely to see growth in the form of the adoption of laser and microwave and other kinetic measures, but at least currently they offer only limited defence against such threats. A great ramification of unmanned systems with cross-domain kill webs is likely be the introduction of joint all-domain command and control of manned and unmanned assets where autonomy would be central to the conduct of operations as control purely by human operators will likely lack the speed to respond to fast developing threats in mission time and acting upon them. This setup will bring together autonomy on weapons, platforms, and the network together completing the digitization process of the battlespace. Particularly with the introduction of large swarms the balance is going to overwhelmingly tip in favour of autonomous networks adjudicating decisions regarding choice of targets, platforms, and weapons. As each state acts to find their own answers to balancing autonomy and the need for human decision-making, ensuring that hostilities remain calibrated in scope and action may require communications around best practices between states on how this balance should shape up. This will be a critical factor in stabilizing and avoiding miscalculations in future crisis situations.

Notes

1 "Affordable Air Power in the era of Gins versus Butter", Talk delivered by Prof. Amit Gupta at Observer Research Foundation, 1 July 2019.
2 "World of Drones", New America, 2019, https://www.newamerica.org/international-security/reports/world-drones/introduction-how-we-became-a-world-of-drones/
3 Ibid.
4 Axe David, "Russia's Predator Drone Flew Strikes in Syria", The National Interest, November 2, 2019, https://nationalinterest.org/blog/buzz/russia%E2%80%99s-predator-drone-flew-strikes-syria-93366
5 Ibid.
6 N.C. Bipindra & Bloomberg News, "Modi Shares Secrets of Predator-like Drone to Bolster India", Chicago Tribune, September 24, 2015, https://www.chicagotribune.com/news/sns-wp-blm-news-bc-india-drones24-20150924-story.html

7 "World of Drones", New America, 2019, https://www.newamerica.org/international-security/reports/world-drones/introduction-how-we-became-a-world-of-drones
8 Ibid.
9 Lin, Jeffrey, "These Are China's New Underwater Drones", Popular Science, March 18, 2019, https://www.popsci.com/finding-qianlong-iii-chinas-new-underwater-drones/
10 Majumdar, Dave, "The U.S. Navy's Robotic Undersea Future", The National Interest, November 3, 2015, https://nationalinterest.org/blog/the-buzz/the-us-navys-robotic-undersea-future-14239
11 Mizokami Kyle, "How the U.S. Air Force Turns an F-16 Fighter into a Drone", Popular Mechanics, November 19, 2019, https://www.popularmechanics.com/military/aviation/a29847417/f-16-drone/
12 Peter Warren Singer. "Wired for War: The Robotics Revolution and Conflict in the 21st Century". Penguin, 2009, pg. 254.
13 Ibid, pg. 224.
14 Ibid, pg. 225.
15 Ibid, pg. 225.
16 Ibid, pg. 230.
17 Ibid, pg. 230.
18 Parsons Dan, "28-Drone Swarm Just Led the Way for a Simulated Air Assault Mission", The Drive, May 24, 2022, https://www.thedrive.com/the-war-zone/28-drone-swarm-just-paved-way-for-simulated-army-air-assault-mission?mc_cid=1ee4eecadd&mc_eid=cb69914e13
19 Peter Warren Singer. "Wired for War: The robotics Revolution and Conflict in the 21st Century". Penguin, 2009, pg. 234.
20 Ibid, pg. 234.
21 Ibid, pg. 235.
22 Suciu Peter, "Does the Nagorno-Karabakh Conflict Prove the Tank Is Toast?", The National Interest, October 5, 2020, https://nationalinterest.org/blog/buzz/does-nagorno-karabakh-conflict-prove-tank-toast-170155
23 Ibid.
24 Cited in Dixon Robyn. "Azerbaijan's Drones Owned the Battlefield in Nagorno-Karabakh and Showed Future of Warfare", The Washington Post, November 11, 2020, https://www.washingtonpost.com/world/europe/nagorno-karabkah-drones-azerbaijan-aremenia/2020/11/11/441bcbd2-193d-11eb-8bda-814ca56e138b_story.html
25 Bateman Robert, "Nagorno-Karabakh Losses Don't Prove the Death of the Tank", Foreign Policy, October 15, 2020, https://foreignpolicy.com/2020/10/15/drones-tanks-obsolete-nagorno-karabakh-azerbaijan-armenia/
26 Ibid.
27 Ibid.
28 Suciu Peter, "Does the Nagorno-Karabakh Conflict Prove the Tank Is Toast?", The National Interest, October 5, 2020, https://nationalinterest.org/blog/buzz/does-nagorno-karabakh-conflict-prove-tank-toast-170155
29 Dixon Robyn, "Azerbaijan's Drones Owned the Battlefield in Nagorno-Karabakh and Showed Future of Warfare", The Washington Post, November 11, 2020, https://www.washingtonpost.com/world/europe/nagorno-karabkah-drones-azerbaijan-aremenia/2020/11/11/441bcbd2-193d-11eb-8bda-814ca56e138b_story.html
30 Axe David, "The U.S. Marines Wants to Dump All Tanks in 10 Years", The National Interest, March 23, 2020. https://nationalinterest.org/blog/buzz/us-marines-wants-dump-all-tanks-10-years-136342
31 "India-China Border Standoff: Heavyweight Tanks and Armoured Vehicles Deployed in Eastern Ladakh", The Financial Express, September 27, 2020, https://

www.financialexpress.com/defence/india-china-border-standoff-heavyweight-tanks-and-armoured-vehicles-deployed-in-eastern-ladakh/2092886/

32 Watling Jack, "The Key to Armenia's Tank Losses: The Sensors, Not the Shooters", Royal United Service Institute Defence Systems, October 6, 2020, https://rusi.org/publication/rusi-defence-systems/key-armenia-tank-losses-sensors-not-shooters

33 Ibid.

34 Ibid.

35 Ibid.

36 Kofman Michael, "The Second Nagorno-Karabakh War, Two Weeks in", War on the Rocks, October 14, 2020. https://warontherocks.com/2020/10/the-second-nagorno-karabakh-war-two-weeks-in/

37 Impact of sub-surface threats are not discussed here because they have been part of the threat equation for long for the surface fleet and are accounted for already in the battlegroup composition.

38 Khurana Gurpreet S, "Aircraft Carriers and India's Naval Doctrine", Manohar Parrikar Institute of Defence Studies and Analysis, 2:1, 2008, https://idsa.in/jds/2_1_2008_AircraftCarriersandIndiaNavalDoctrine_GSKhurana

39 Ibid.

40 Report, "Military and Security Developments Involving the People's Republic of China", Department of Defense, 2020, https://media.defense.gov/2020/Sep/01/2002488689/-1/-1/1/2020-DOD-CHINA-MILITARY-POWER-REPORT-FINAL.PDF. pg. 56.

41 Thompson Loren, "Why China Can't Target U.S. Aircraft Carriers", Forbes, August 9, 2019, https://www.forbes.com/sites/lorenthompson/2019/08/09/why-china-cant-target-u-s-aircraft-carriers/?sh=5d1d288f716a

42 Ibid.

43 Ibid.

44 Ibid.

45 La Grone, Sam, "Report: U.S. Carriers Need New Lethal Unmanned Aircraft, New Fighter to Stay Relevant", US Naval Institute, March 5, 2019, https://news.usni.org/2019/03/05/report-u-s-carriers-need-new-lethal-unmanned-aircraft-new-fighter-to-stay-relevant

46 "U.S. Navy Awards $1B in Contracts for Unmanned Surface Vessel Program", The Maritime Executive, February 19, 2021, https://www.maritime-executive.com/article/u-s-navy-awards-1b-in-contracts-for-unmanned-surface-vehicle-systems

47 Shelbourne Mallory, "Navy to Use Sea Hunter in Fleet Exercises as Unmanned Systems Experimentation Continues", US Naval Institute, September 30, 2020, https://news.usni.org/2020/09/30/navy-to-use-sea-hunter-in-fleet-exercises-as-unmanned-systems-experimentation-continues

48 Sutton HI, "Chinese HSU-001 LDUUV Large Displacement Unmanned Underwater Vehicle", hisutton.com, October 2, 2019, http://www.hisutton.com/Chinese_LDUUV.html

49 Singh Abhijit, "How India, too, Is on a Quest for Undersea Dominance, to Counter the Chinese Navy's Growing Presence", Observer Research Foundation, August 31, 2019, https://www.orfonline.org/research/43742-how-india-too-is-on-a-quest-for-undersea-dominance-to-counter-the-chinese-navys-growing-presence/

50 Ibid.

51 Ibid.

52 Davis Malcom, "Robots at War: The Future for Autonomous Systems at Sea in the Indo-Pacific", Observer Research Foundation, March 24, 2021, https://www.orfonline.org/expert-speak/robots-war-future-autonomous-systems-sea-indo-pacific/?amp#_ednref12

53 Bajema Natasha, "Will AI Steal Submarines' Stealth?", IEEE Spectrum, July 16, 2022, https://spectrum.ieee.org/nuclear-submarine

54 Singh Abhijit, "How India, too, Is on a Quest for Undersea Dominance, to Counter the Chinese Navy's Growing Presence", Observer Research Foundation, August 31, 2019, https://www.orfonline.org/research/43742-how-india-too-is-on-a-quest-for-undersea-dominance-to-counter-the-chinese-navys-growing-presence/
55 "Navy Large Unmanned Surface and Undersea Vehicles: Background and Issues for Congress", Congressional Research Service, March 17, 2021, https://fas.org/sgp/crs/weapons/R45757.pdf
56 Trevithick, Joseph, "The Navy Plans to Launch Swarms of Aerial Drones from Unmanned Submarines and Ships", The Drive, March 1, 2021, https://www.thedrive.com/the-war-zone/39535/navy-contract-exposes-plans-to-launch-swarms-of-drones-from-unmanned-boats-and-submarines
57 Footnoted in "Navy Large Unmanned Surface and Undersea Vehicles: Background and Issues for Congress", Congressional Research Service, March 17, 2021, https://fas.org/sgp/crs/weapons/R45757.pdf. pg. 5.
58 Cited in ibid, pg. 6.
59 Cenciotti David, "Let's Forget the Hype and Talk about the AI vs Human Pilot Simulated Dogfight", The Aviationist, August 24, 2020, https://theaviationist.com/2020/08/24/lets-forget-the-hype-and-talk-about-the-ai-vs-human-pilot-simulated-dogfight/
60 Ibid.
61 Tucker Patrick, "An AI Just Beat a Human F-16 Pilot in a Dogfight — Again", Defense One, August 20, 2020, https://www.defenseone.com/technology/2020/08/ai-just-beat-human-f-16-pilot-dogfight-again/167872/
62 Ibid.
63 Cenciotti David, "Let's Forget the Hype and Talk about the AI vs Human Pilot Simulated Dogfight", The Aviationist, August 24, 2020, https://theaviationist.com/2020/08/24/lets-forget-the-hype-and-talk-about-the-ai-vs-human-pilot-simulated-dogfight/
64 Ibid.
65 Tucker Patrick, "An AI Just Beat a Human F-16 Pilot in a Dogfight — Again", Defense One, August 20, 2020, https://www.defenseone.com/technology/2020/08/ai-just-beat-human-f-16-pilot-dogfight-again/167872/
66 Cenciotti David, "Let's Forget the Hype and Talk about the AI vs Human Pilot Simulated Dogfight", The Aviationist, August 24, 2020, https://theaviationist.com/2020/08/24/lets-forget-the-hype-and-talk-about-the-ai-vs-human-pilot-simulated-dogfight/
67 Axe David, "Russia's Got Dogfighting Drones, Too—Just How Smart Are They?", Forbes, August 29, 2020, https://www.forbes.com/sites/davidaxe/2020/08/31/russias-got-dogfighting-drones-too-just-how-smart-are-they/
68 Tucker Patrick, "Russia Is Working to Pair Combat Jets and Drones, Too", Defense One, February 17, 2021, https://www.defenseone.com/technology/2021/02/russia-working-pair-combat-jets-and-drones-too/172109/
69 "DARPA Initiates Design of LongShot Unmanned Air Vehicle", DARPA, August 2, 2021, https://www.darpa.mil/news-events/2021-02-08
70 Florian. "Human-Robot Interactions in Future Military Operations". CRC Press, 2016. pg. 120.
71 Ibid, pg. 108–124, 162.
72 Ibid, pg. 158–159.
73 Cited in John Gen P. Abizaid, Ret Army & Rosa Brooks. "The Task Force on US Drone Policy Co-Chairs Second Edition", Stimson Center April, 2015, https://www.stimson.org/wp-content/files/file-attachments/recommendations_and_report_of_the_task_force_on_us_drone_policy_second_edition.pdf. pg. 22.
74 Journal of Strategic Security, Vol. 7, No. 4, Special Issue: Future Challenges in Drone Geopolitics (Winter 2014), https://www.jstor.org/stable/pdf/26465233.pdf pg. 115.

75 Singer Peter Warren. "Wired for War: The Robotics Revolution and Conflict in the 21st Century". Penguin, 2009. pg. 199.
76 Ibid, pg. 199.
77 Ibid, pg. 199.
78 Peterson Scott & Payam Faramarzi, "Exclusive: Iran Hijacked US Drone, Says Iranian Engineer", The Christian Science Monitor, December 15, 2011, https://www.csmonitor.com/World/Middle-East/2011/1215/Exclusive-Iran-hijacked-US-drone-says-Iranian-engineer
79 Cited in John Gen P Abizaid, Ret Army & Rosa Brooks, "The Task Force on US Drone Policy Co-Chairs Second Edition", Stimson Center April, 2015, https://www.stimson.org/wp-content/files/file-attachments/recommendations_and_report_of_the_task_force_on_us_drone_policy_second_edition.pdf. pg. 22.
80 Cited in Johnson James S. (2020) Artificial intelligence: A threat to strategic stability, Strategic Studies Quarterly, 14:1, 16–39, 2021, DOI: 10.2307/26891882. pg. 20.
81 Peter Warren Singer. "Wired for War: The Robotics Revolution and Conflict in the 21st Century". Penguin, 2009. pg. 200.
82 Ibid, pg. 202.
83 McLeary Paul, "US Warship Fries Drone with Powerful New Laser", Breaking Defense, May 22, 2020, https://breakingdefense.com/2020/05/us-warship-fries-drone-with-powerful-new-laser/
84 Hitchens Theresa & Sydney J. Freedberg, "Troops to Test AFRL's THOR Drone Killer This Fall", Breaking Defense, August 4, 2020, https://breakingdefense.com/2020/08/troops-in-theater-to-test-afrls-thor-drone-killer-this-fall/
85 Ibid.
86 Ibid.
87 Nagel Jacob, Bradley Bowman & Maj. Liane Zivitski, "Assessing Israel's Tactical Laser Breakthrough", Defense News, January 17, 2020, https://www.defensenews.com/opinion/commentary/2020/01/17/how-realistic-is-israels-tactical-laser-breakthrough/
88 Ibid.
89 Rogoway Tyler, "We Talk Suicide Drones and the Future of Unmanned Warfare with AeroVironment's Steve Gitlin", The Drive, July 8, 2020, https://www.thedrive.com/the-war-zone/34414/we-talk-killer-drones-and-the-future-of-unmanned-warfare-with-aerovironments-steve-gitlin
90 Trevithick Joseph, "The Navy Plans to Launch Swarms of Aerial Drones from Unmanned Submarines and Ships", The Drive, March 2, 2021, https://www.thedrive.com/the-war-zone/39535/navy-contract-exposes-plans-to-launch-swarms-of-drones-from-unmanned-boats-and-submarines
91 Watling Jack, "The Key to Armenia's Tank Losses: The Sensors, Not the Shooters", Royal United Service Institute, October 6, 2020, https://rusi.org/publication/rusi-defence-systems/key-armenia-tank-losses-sensors-not-shooters
92 Kofman Michael, "The Second Nagorno-Karabakh War, Two Weeks in", War on the Rocks, October 14, 2020, https://warontherocks.com/2020/10/the-second-nagorno-karabakh-war-two-weeks-in/
93 Davis Malcom, "Robots at War: The Future for Autonomous Systems at Sea in the Indo-Pacific", Observer Research Foundation, March 24, 2021, https://www.orfonline.org/expert-speak/robots-war-future-autonomous-systems-sea-indo-pacific/?amp#_edn4
94 Ibid.
95 Ibid.
96 Kube, Courtney, Phil Helsel & Ali Arouzi, "U.S. Drone Shot down by Iran in International Space, U.S. Officials Say", NBC News, June 20, 2019, https://www.nbcnews.com/news/world/military-official-says-no-u-s-drones-iranian-airspace-after-n1019566

97 "Qasem Soleimani: US Kills Top Iranian General in Baghdad Air Strike", BBC News, January 3, 2020, https://www.bbc.com/news/world-middle-east-50979463
98 "Iran Plane Crash: Ukrainian Jet Was 'Unintentionally' Shot Down", BBC News, January 11, 2020, https://www.bbc.com/news/world-middle-east-51073621
99 Cited in Jentsch Florian. "Human-Robot Interactions in Future Military Operations". CRC Press, 2016. pg. 72.
100 Kott Alexander, ed. "Battle of Cognition: The Future Information-rich Warfare and the Mind of the Commander". Greenwood Publishing Group, 2008, pg. 32.
101 Field Manual (FM) 3-0, Operations, October 2017, Chapter -1, pg. 29.
102 Florian, "Human-Robot Interactions in Future Military Operations". CRC Press, 2016. pg. 74.
103 Pravin Sahwney, "Why India Must Avoid Hitching Itself to US Military's Plans for China and the Indo-Pacific", The Wire, 2018, https://thewire.in/security/why-india-must-avoid-hitching-itself-to-us-militarys-plans-for-china-and-the-indo-pacific
104 Field Manual (FM) 3-0, Operations, October 2017, Chapter -1, pg. 39.
105 Ibid., pg. 34.
106 Kott Alexander, ed. "Battle of Cognition: The Future Information-rich Warfare and the Mind of the Commander". Greenwood Publishing Group, 2008. pg. 30–31.
107 Ibid, pg. 31.
108 Paolo Spagnoletti & Andrea Salvi, "Digitization of Mission Command and Control: Fragmenting Coordination in Military Operations" in NATO Decision-making in the Age of Big Data and Artificial Intelligence, ed. Sonia Lucarelli, Allesandro Marrone & Francesco N. Moro, 2021, https://www.iai.it/sites/default/files/978195445000.pdf. pg. 28.
109 Field Manual (FM) 3-0, Operations, October 2017, Chapter -1, pg. 17.
110 Kott Alexander, ed. "Battle of Cognition: The Future Information-rich Warfare and the Mind of the Commander". Greenwood Publishing Group, 2008. pg. 30.
111 Crane Conrad, "Mission Command and Multi-Domain Battle Don't Mix", War on the Rocks, August 23, 2017, https://warontherocks.com/2017/08/mission-command-and-multi-domain-battle-dont-mix/
112 Schneider Jacquelyn & Julia Macdonald, "Unmanned Capabilities and the Future of Great Power Competition and Conflict", Social Science Research Network, electronic Journal, 2019, https://doi.org/10.2139/ssrn.3407957
113 Peter Warren Singer. "Wired for War: The Robotics Revolution and Conflict in the 21st Century". Penguin, 2009, pg. 222.
114 Strout Nathan, "Inside the Army's Futuristic Test of Its Battlefield Artificial Intelligence in the Desert", C4ISRNET, September 25, 2020, https://www.c4isrnet.com/artificial-intelligence/2020/09/25/the-army-just-conducted-a-massive-test-of-its-battlefield-artificial-intelligence-in-the-desert/
115 Osborn Kris, "Copy Cat: Even China Wants a Highly-Networked Military", The National Interest, January 27, 2021, https://nationalinterest.org/blog/buzz/copy-cat-even-china-wants-highly-networked-military-177160
116 "Joint All Domain Command and Control", Congressional Research Service, March 18, 2021, https://fas.org/sgp/crs/natsec/IF11493.pdf
117 Niewood Eliahu, Greg Grant & Tyler Lewis, "A NEW BATTLE COMMAND ARCHITECTURE for MULTI-DOMAIN OPERATIONS Countering Peer Adversary Power Projection", https://www.mitre.org/sites/default/files/publications/Joint-All-Domain-Command-Control.pdf
118 Ibid, pg. 4.
119 Henrik Røboe Dam, "Central to Meeting the Complexities of JADC2? Artificial Intelligence", Defense News, May 13, 2020, https://www.defensenews.com/2020/05/13/central-to-meeting-the-complexities-of-jadc2-artificial-intelligence/

120 Konaev Margarita, "With AI, We'll See Faster Fights, but Longer Wars", War on the Rocks, October 29, 2019, https://warontherocks.com/2019/10/with-ai-well-see-faster-fights-but-longer-wars/
121 "Long Range Anti-Ship Missile (LRASM)", Lockheed Martin, 2018, https://www.lockheedmartin.com/en-us/products/long-range-anti-ship-missile.html
122 Davis Zachary, "Artificial Intelligence on the Battlefield: An Initial Survey of Potential Implications for Deterrence, Stability, and Strategic Surprise", March 2019, https://cgsr.llnl.gov/content/assets/docs/CGSR-AI_BattlefieldWEB.pdf. pg. 6–7.
123 Cited in ibid, pg. 7.
124 Kott Alexander, ed. "Battle of Cognition: The Future Information-rich Warfare and the Mind of the Commander". Greenwood Publishing Group, 2008. pg. 20.
125 Cited in John Gen P Abizaid, Ret Army & Rosa Brooks. "The Task Force on US Drone Policy Co-Chairs Second Edition", Stimson Center April, 2015, https://www.stimson.org/wp-content/files/file-attachments/recommendations_and_report_of_the_task_force_on_us_drone_policy_second_edition.pdf. pg. 26.
126 Kott Alexander, ed. "Battle of Cognition: The Future Information-rich Warfare and the Mind of the Commander". Greenwood Publishing Group, 2008. pg. 24–25.
127 Ibid, pg. 26–27.
128 Peter Warren Singer. "Wired for War: The Robotics Revolution and Conflict in the 21st Century". Penguin, 2009. pg. 350.
129 Ibid, pg. 352.
130 Henrik Røboe Dam, "Central to Meeting the Complexities of JADC2? Artificial Intelligence", Defense News, May 13, 2020, https://www.defensenews.com/2020/05/13/central-to-meeting-the-complexities-of-jadc2-artificial-intelligence/
131 Ibid, pg. 20.
132 Cited in Solen Derek, "Chinese Views of All-Domain Operations", China Aerospace Studies Institute, August 2020, https://www.airuniversity.af.edu/Portals/10/CASI/documents/Research/CASI%20Articles/2020-06-30%20Chinese%20Views%20of%20All-Domain%20Operations.pdf
133 Cited in Jentsch Florian. "Human-Robot Interactions in Future Military Operations". CRC Press, 2016. pg. 69.
134 Ibid, pg. 200.
135 Peter Warren Singer. "Wired for War: The Robotics Revolution and Conflict in the 21st Century". Penguin, 2009. pg. 357.
136 John Gen P Abizaid, Ret Army & Rosa Brooks, "The Task Force on US Drone Policy Co-Chairs Second Edition", Stimson Center, April, 2015, https://www.stimson.org/wp-content/files/file-attachments/recommendations_and_report_of_the_task_force_on_us_drone_policy_second_edition.pdf. pg. 26.
137 Davis Zachary, "Artificial Intelligence on the Battlefield: An Initial Survey of Potential Implications for Deterrence, Stability, and Strategic Surprise", Center for Global Security Research, March 2019, https://cgsr.llnl.gov/content/assets/docs/CGSR-AI_BattlefieldWEB.pdf. pg. 7–8, 9.
138 The technical feasibility of this claim is contested. See Johnson James S. (2020) Artificial intelligence: A threat to strategic stability, Strategic Studies Quarterly, 14:1, 16–39, DOI: 10.2307/26891882
139 Peter Warren Singer. "Wired for War: The Robotics Revolution and Conflict in the 21st Century". Penguin, 2009. pg. 226.
140 Ibid, pg. 226.
141 Johnson James S. (2020) Artificial intelligence: A threat to strategic stability, Strategic Studies Quarterly, 14:1, 16–39, Doi: 10.2307/26891882. pg. 22.
142 Ibid, pg. 19.
143 Kallenborn Zachary, "Drones of Mass Destruction: Drone Swarms and the Future of Nuclear, Chemical, and Biological Weapons", War on the Rocks, February 14,

2019, https://warontherocks.com/2019/02/drones-of-mass-destruction-drone-swarms-and-the-future-of-nuclear-chemical-and-biological-weapons/

144 Leslie F. Hauck III & Dr. John P. Geis II, "Air Mines Countering the Drone Threat to Aircraft", Air & Space Power, Journalhttps://www.airuniversity.af.edu/Portals/10/ASPJ/journals/Volume-31_Issue-1/V-Hauck_Geis.pdf

145 Kallenborn Zachary, "Drones of Mass Destruction: Drone Swarms and the Future of Nuclear, Chemical, and Biological Weapons", War on the Rocks, February 14, 2019, https://warontherocks.com/2019/02/drones-of-mass-destruction-drone-swarms-and-the-future-of-nuclear-chemical-and-biological-weapons/

146 Ibid, pg. 196.

147 Ibid, pg. 196.

148 Ibid, pg. 196.

149 Ibid, pg. 196.

150 Numerous more such instances have been compiled by Peter Singer in his book *Wired for War.*

4

OFFENCE-DEFENCE INTEGRATION AND NUCLEAR AMBIGUITY

Effect on Conventional Stability

The third nuclear age is essentially the net impact of its key characteristics, namely the introduction of new military technologies and the quantitative and qualitative expansion of the nuclear arsenal by the states. These trends do not exist independently in the conflict spectrum and have a relationship with conventional conflict and strategy. The combination of expanding nuclear arsenals with emerging technologies in the conventional space and newer strategies to shape the overall conventional conflict and therefore conventional deterrence all contribute to establishing relationships between different levels of the conflict spectrum in the third nuclear age which act in unison. This chapter delves into these relationships and their impact on stability in the conventional realm, in terms of an essential concept of offence-defence balance in the spectrum. Technologies like hypersonic missiles, unmanned vehicles, cyber or electronic warfare along with concepts like kill web and conventional prompt global strikes seem to add to the "strategic complexity" that already exists. Their net outcome shapes stability or instability that already exists in the conventional realm. As described in the first chapter, inadvertent and advertent routes to nuclear escalation exist in the conflict spectrum and we define that space as the conventional-nuclear intersection. These routes originate in both technology and strategy with their net impact pressuring states to either keep out of that space altogether or form strategies to take advantage of this conventional-nuclear intersection and optimize their current and emerging weapons and strategies to aid their conventional objectives. The use of escalation prospects in the conventional-nuclear intersection space towards shaping conventional war outcomes is a fundamental aspect of instrumentalized conventional strategy and key to understanding stability concerns expressed in the third nuclear age. These include the ramifications of qualitative and quantitative expansion of the nuclear arsenals that in fact carry over to the conventional conflict spectrum. Doctrine,

DOI: 10.4324/9781003365976-5

strategy, and technology in effect combine to establish relationships between different levels in the conflict spectrum that make instrumentalized conventional strategy more potent. This chapter starts with a discussion of the offence-defence theory and subsequently applies it to each level of the conventional conflict to judge the offence or defence dominance at that level. This offence-defence balance within the concept of instrumentalized conventional strategy impacts the peacetime coercion stage, escalation control, or intra-war deterrence stage and finally the conventional-nuclear transition point in the conflict spectrum that are analogous to tactical, operational, and strategic levels within conventional operations. A relationship therefore exists in the form of an offence-defence integration across levels of the conventional operations rather than systems and allows for higher-level capabilities to be used to address issues in the conventional realm.

Offence-defence balance theory

At its heart "the offense-defense theory has two key variables one is the offense-defense balance" that refers to one the relative ease in attacking or defending against targets[1] and the second "offence-defence distinguishability that refers to whether forces that support offensive missions are different from those that support defensive missions". This concept can be useful to sum up the impact of a variety of developments in the technological domain and relate them to the conventional conflict spectrum. Glaser and Kaufmann define the offence-defence balance as a key variable along with the power that determines the state's ability to perform military missions. In general the more expensive forces can afford to take territory at lower cost than less expensive forces; this is the minimum investment ratio at which the attacker can take territory at an acceptable cost of fighting.[2] Simply, when the offence-defence balance is more in favour of offence, then the security dilemma between countries becomes more acute and it also results in arms racing thus leading to overall instability in the system.[3] On the contrary if a more defence-dominant balance exists, then the system is likely to be more stable because sufficient defensive capability can be attained far cheaply for defensive purposes than offensive capability for offensive missions.

Military goals have a direct bearing on this balance, and revision in territorial goals is likely to shift the offence-defence balance. Given a finite defensive force, the size and skill of the offensive forces required for the mission will vary with the mission's aim; a more ambitious territorial goal will naturally require more troops with higher level of technology.[4] An ambitious military goal will also require deeper penetration by military forces along with more vulnerable and longer supply lines over often a protracted timeline giving defence an edge.[5] This makes military missions with ambitious goals naturally defence-dominant.

Glasser and Kaufmann also hold the view that the overall offence-defence balance should be defined at the strategic level since the outcome of the war is determined by the final outcome and being able to hold territory temporarily does not imply an offensive advantage since counter-attacks can reverse those territorial gains.[6] But they do admit that to gain territory the attacker must be able to conduct offensive tactical battles and offensive operations, and similarly to hold territory they must be able to successfully defend at the tactical or the operational level. They argue that since any strategic offensive requires offensive operations and they in turn require offensive tactical battles, a change that makes tactical offensive easier or harder is likely to do the same for operational-level offence and hence strategic level offence.[7]

Finally offence-defence theory assumes optimality; that is, military doctrines and forces of states cannot influence the objective balance as doctrine and deployments are merely outcomes of the optimal posture at a given limitation of resources and conditions imposed by the offence-defence balance.[8] Assuming optimal behaviour allows an objective assessment of the overall impact that introduction of new weapons systems and technologies particularly missiles, unmanned systems, and other technologies might have on the stability of the overall system and in a particular dyad as different states will eventually employ them in the most optimal way.

There are three primary ways in which offence-defence balance could be altered: one is the introduction of a new form of conflict with a distinct offence-defence dynamics; the second could be introduction of a new technology or weaponry that dramatically alters the relative force size required for offence or defence; and finally if the investment in offensive or defensive forces becomes dramatically more manageable on account of economic growth or increasing cost effectiveness or power of weapons and technologies used in battle.[9] The first two are more relevant in terms of the arrival of the third nuclear age and its emerging doctrines and technologies, and the discussion in the text later will bring that aspect out at the tactical, operational, and strategic levels of the conventional conflict.

Offence-dominance and instrumentalized conventional strategy

There are two direct consequences of the offence-defence theory, one that if the military goal is modest or relatively small, the input cost of conducting an aggressive military action is negligible to the aggressor. Most local or tactical situations that present relatively lower-hanging military challenges are therefore likely to be a case where offence is dominant. Second, offence dominance at the tactical level is likely to make conflict at the operational level or strategic level also offence-dominant or easier for offensive military strategy at those levels. Lower input cost is particularly of direct consequence to a peacetime coercive strategy that calibrates military goals, thereby inherently promoting an offence-dominant offence-defence balance in the system. The calibrated military goals don't require very expensive forces; often the existing forces just adopt a military posture that reinforces tactically offensive and operationally and strategically defensive stance accomplishing a fait accompli and leading to a territorial gain even without fight, thus making those gains without the exorbitant cost of fighting. Due to this offence dominance at the tactical level fait accompli is naturally a more preferred choice of strategy that is used keeping in perspective adversary threshold levels and own power asymmetry. It is, however, possible that a limited territorial aim strategy may in fact aid defence if it allows the defender to regroup and retaliate, in which case a total war for total defeat may be more suited goal for an offensive strategy.[10] But instrumentalized conventional strategy that critically works on the principle of mutual reinforcement by using the newly acquired military gains to further consolidate the military situation on the ground attempts to negate the regrouping advantage by resorting to an intra-war deterrent stance at the operational level. Regardless of the offence-defence balance at the operational or strategic levels, the overall object of instrumentalized conventional strategy is therefore offence-dominant, as was also brought out in the first chapter.

Second in an instrumentalized conventional strategy where calibrated gains are made and these gains are in fact mutually reinforcing with the final military posture attained by the move, a tactical gain is quickly converted into an operational defence task or objective by the aggressor. At this new operational-level intra-war deterrence and escalation control is utilized. As discussed earlier, the balance of resolve does not aid the defender here who must wage a battle at the operational level for essentially a tactical loss. Unlike the case propounded by the offence-defence theory that offence-dominant tactical action would make operational and strategic levels also offence-dominant, in an instrumentalized conventional strategy offence at the tactical level is followed by defence at the operational and strategic levels.

The impact of these factors in overall conventional deterrence is pronounced. In essence the local balance of forces is critical in determining the stability or instability of a deterrent equation.[11] They impact the aggressors' calculation about achieving a rapid victory, thereby affecting the offence-defence balance.[12] This also means that overall conventional deterrence may get affected if there is a failure in local-level deterrence despite an overall balance of power. In a sense conventional deterrence is then an outcome or multiple separate and unique local-level deterrence equations between nuclear adversaries. This is a very key feature of the third nuclear age where strenuous offence-defence balance at the local level gives an overall unstable character to the conventional deterrence equations between nuclear weapon states. The fact that these local-level deterrence equations are very much a function of opportunity in space and time means that overall conventional deterrence is never stable. Consequentially "conventional deterrence puts a premium on forward-deployed combat power, as well as forcible entry, force sustainment and reinforcement capabilities, and regional base access".[13]

The doctrine of "local wars under high tech conditions" is therefore a general principle that is borne out in offence-defence balance in the conventional realm making instability a unique feature of the third nuclear age and clashes between nuclear armed states appear more feasible. The reason feasibility of nuclear crises appears more prominent and in fact why every conventional crisis is tinged with a nuclear overhang is that there are enough conditions under which this instability may lead to escalation or find its way to the conventional-nuclear transition point which is discussed in the subsequent sections.

There is, however, more to the instability derived from local offence-defence balance. As this instability is a function in time-space, it can be periodic in nature. Less frequent or more frequent but frequent nevertheless, this periodicity probes deterrence thresholds between nuclear weapon states in a gradual manner. These are likely to lead to iterative crises where deterrence equations are shaped over multiple iterations rather than the outcome of a single engagement in a one-off episode. On one hand these might allow states to stay within the conventional realm given the confidence and the expectation that prior behaviour will be replicated in some manner. The scope of targets and the nature of targets may only see a well-thought-out calibration from both sides to ensure iterative engagements do not translate into large-scale engagement of forces. Yet, over time they are more likely to push the envelope for states on what means could be employed and what could be targeted in a particular deterrence dyad, thus shifting the thresholds towards the higher side on the escalation ladder. These are likely to have second-order effects in judging a state's political resolve over time and generate higher probability of miscalculation in anticipating adversary reaction during conventional crises. It may be difficult to predict behaviour in dyads at a higher number of iterations when the space to manoeuvre below the threshold is reduced and eventually it is crossed. However other factors like how spaced out these iterations may be and public perception related to thresholds may have a stabilizing effect, leading to frequent episodes of bounded escalation. In addition, iterative crisis may allow for a variety of smaller-scale actions to be dressed as punitive options that can then be used as deterrence by punishment options, thereby enhancing crisis stability as the damage inflicted and received is within tolerable limits or could be shown to be within tolerable limits justifying no military counter-response.

Two cases in the third nuclear age are pertinent here that are useful to expound on the implications of the above, particularly the offence dominance at the tactical level under an instrumentalized conventional strategy. One is the progression of crises in the deterrence dyad between India and Pakistan. Iterative deterrence equation between India and Pakistan saw military responses in 2016 and 2019. In 2016, a militant attack conducted by a militant outfit inflicted double-digit fatalities on Indian security forces which was a breach threshold as for a long time no armed clash had taken place in the country where such a high number of fatalities was caused. India responded with a tactical military operation across the Line of Control (LoC) which, though had happened previously, had important differences. For the first time a military operation was conducted and publicly acknowledged; unlike previous military cross-border raids, this was the first time the action was conducted across multiple points giving a relatively larger scale than had been achieved in previous such cross-border actions. Pakistan stated that no such military action was conducted, and this allowed India to get away without any kind of retaliatory military action. This iteration, however, established a template of publicly acknowledged military response to a militant clash that leads to high double-digit fatality event as a means of "achieving" or "re-establishing" deterrence at the tactical level. This was followed by a second iteration of the crisis in 2019 when a suicide attack conducted and claiming affiliation with Pakistan-based Jaish-e-Mohammad (JeM) inflicted 40 fatalities on an Indian paramilitary unit, the highest death toll in a single incident in the former state of Jammu and Kashmir and in fact in any militant attack against Indian security forces. Having entered an iterative deterrence equation with Pakistan, this time India conducted Balakot airstrikes, in mainland territory of Pakistan, on a militant target in retaliation. This time Pakistan counter escalated with its own air strike, and this ended in duel between Indian and Pakistan fighter jets, a first since the 1971 full-fledged conventional war, leading to the shootdown of one Indian fighter jet. Not only did this iteration cement a military response template for both sides, it moved the military interaction a few thresholds up the escalation ladder. One vertical escalation by India, use of fighter planes, was responded to in equal measure by Pakistan, even though in both iterations India hit militant targets and not regular military targets. India crossed spatial thresholds in its attack from retaliating within the Kashmir theatre in 2016 to attacking in mainland Khyber Pahktunkhwa in 2019. However, in both Pakistan's and India's cases the aerial ingresses were made across the LoC and not over international boundary, suggesting that in a future iteration this threshold might be adhered to by both sides.[14] Overall, these airstrikes added to the credibility of that threshold just as much as they were a result of the threshold that had already been built. The fact that these events are spaced out may allow room for calibration downwards in the next iteration, but simultaneously, they also evince how the conflict envelope has been pushed forward and the next iteration could see pressure to adhere to response templates creating a pressure to escalate further upwards. The cumulative effects of iterative crises, especially if they are not spaced out in time, could push the crises higher on the escalation ladder.

A second is the India-China clash in Galwan, Ladakh, in 2020. In this instance Chinese forces were able to take advantage of a pandemic that delayed an Indian military deployment to implement a military fait accompli across multiple points of intrusion all along the Line of Actual Control (LAC), and during an effort to disengage a confrontation ensued, in which 20 Indian soldiers were killed and 56 others were left injured while an unknown number of Chinese soldiers were also killed.[15] The tactical gains that the Chinese made allowed them to look over an important tactical road link and interdict it, allowing them to achieve an operational impact in the entire sector by cutting off supplies to an important air base in the area. This tactical offensive was not followed by an operational-level offensive stance; rather, an operationally defensive stance was undertaken with the intent of making the tactical indistinguishable from the operational level and thereby introducing strategic calculations. This was followed by the movement of heavy forces by the Chinese in the rear to shape the military operational-level picture while reinforcing the gains made at the tactical level.[16] In essence the effort is to reduce the distinction between tactical and the operational level once the military gain has fructified. These rear operations that are aimed at the operational level are a key feature in the third nuclear age and the reason why seemingly smaller-level crises eventually almost always take the shape of theatre-level objectives and intra-war deterrence and escalation stages are reached much sooner than anticipated. The relatively lower scale of the tactical offensive is eventually followed by military movements that could potentially lead to enlargement of the scale or intensity of the conflict.

Overall, the offence dominance at the tactical level results in a combination of iterative crises and mingling of the tactical with the operational-level military posture, both of which combine to produce conventional crises that seem to have higher potential than before to take the shape of a major battle and therefore present non-linear paths to the conventional-nuclear transition point. This offence dominance at the tactical level does not allow nuclear armed states to have confidence in deterrence through the overall conventional balance of forces and military power because a breakdown in local-level deterrence effectively opens the door to higher-level instability at the operational and strategic levels. Because of the inherent offence-dominance established through a fait accompli–based military strategy, in effect deterrence is less effective at any point in time-space at lower scales of the conflict.

Defence dominance at the operational level

Feasibility of instrumentalized conventional strategy critically depends on exploitation and stretching defensive dominance at the operational level of the conventional conflict. While offensive dominance at the tactical level is result of the object of the military strategy that is employed by the nuclear-armed states, defence dominance at the operational level is gained through instruments of both strategy and technology. On the technology side a variety of existing, evolving, and emerging capabilities contribute in shaping this defence dominance, making use of peacetime coercion more feasible using the conventional military instrument. These include the operationalization of the kill web, the emerging capabilities in unmanned vehicles, and, most importantly, the existing, but evolving, missile arsenals of nuclear-armed states that shape the operational level of the conflict.

Technology is one of the most important factors influencing the offence-defence balance.[17] The way a particular technology impacts the offence-defence balance is determined by whether or not it adds to a state's ability to perform a defensive or offensive mission; this assessment involves how a particular technology or innovation affects advancing (aggressing) forces and the non-advancing (or defending forces).[18] Technologies that favour non-advancing forces are likely to favour defence, whereas technologies that equally aid advancing forces are likely to favour offence.[19] Glaser and Kaufmann consider six major aspects of technology that are relevant: mobility, firepower, protection, logistics, communication, and detection.[20] Technologies that aid firepower are generally considered to aid defence dominance.[21] To conduct an offensive military mission at the operational level often requires concentration of forces to achieve local superiority and overwhelm the defensive lines of the adversary.[22] But such concentration of forces presents very lucrative targets for defensive firepower particularly area-effect weapons including artillery, missiles carrying conventional warheads, or nuclear warheads that intend to play the purported role of a tactical nuclear weapon.[23] Apart from inflicting damage on the advancing forces they can also take out vital transport links, thus reducing their mobility.[24] However, kill web–enabled technologies have now shown that in the tactical battle area it can lend precision to even artillery, thus making the distinction between defence and offence unclear.

For similar reasons, the most important set of capabilities that afford and enable defence dominance at the operational level of the conflict are therefore missiles.[25] Unlike air power, missiles come at much lesser cost of investment in greater numbers that are expendable with an inevitable damage infliction capability at greater range and at the least amount of risk to the operator. There are exceptions like specific categories of weapons, for example, anti-radiation missiles that are perfect for advancing air forces for breaching areas that are defended with air-defence or missile-defence systems,[26] but in general the variety of missiles and their targets favour usage by defenders and defensive doctrines like A2/AD. They very often play the role of the intra-war deterrent, and their introduction in a theatre gives an operational-level dimension to tactical-level confrontations along with the threat of higher-level escalation.

Missiles appeared much before the third nuclear age, but their contribution to its unique unstable deterrence equations has been a continuing and evolving phenomenon. Though they are potent for both advancing and defending forces, in an instrumentalized conventional strategy, missiles are more likely to aid the first mover or the aggressor who will switch to an operational defensive task once the initial mutually reinforcing military gains are made. They present a higher-level threat on the escalation ladder, and they increase the cost of offensive missions simply by virtue of their potential entry into the conflict. Their longer range, accuracy, and ease to be produced in greater numbers put a large number of vital military (and civilian targets) at risk simply by their potential use or mode of deployment escalating the threat and cost for the aggressor if he chooses to attack. They are also versatile as the same category of missile that can take out targets on land with a different seeker can also very effectively sink ships.[27]

Missile arsenals have also evolved across states. Almost all major nuclear-armed states have moved or begun their transition from using liquid fuel missiles to solid fuel missiles over the last decade.[28] Solid fuel allows canisterization that translates into speedy deployment, mobility, and launch concealment and make them less vulnerable to an adversary hit because there is no need to fuel them before launch.[29] Greater integration of short- and intermediate-ranged missiles that are solid-fuelled, cannesterized, and road- and rail-mobile has happened across all missile-armed states nuclear and non-nuclear and can serve domains ranging from land, sea, and air in all theatres.

The dependency on missiles has in addition provoked the development of missile defence systems that intend to shift the traditional offence-defence balance. As a result, nuclear-armed states have invested in newer kind of missile systems and missile defence systems along with new doctrinal practices. Missile defence systems are, however, typically more expensive and inaccurate to operate. And therefore, the scaling effect of investments in either offensive or defensive systems, in this case missiles and missile defence systems, usually ends in favour of missiles which are the defensive system at the operational level of the conventional war. This is due to the inaccuracy of the missile interceptors that leads to gap exploitation, as when both the aggressor and defender scale up their investments the probability that missiles will breach the defence is more likely initially.[30] Overall, the cheap cost of investment for missile aggressors translates into a greater advantage here as they can simply overwhelm these layered defences with a missile build-up.[31] States have also added cruise missiles in their arsenal which are powered by air breathing engines and can fly at a low altitude.[32] These missiles are difficult to be intercepted by theatre missile defences as they are not always captured by the radar due to their low altitude trajectory.[33] It is even more difficult to detect their firing and launch positions as they can be put on very mobile platforms. Due to these challenges the solution is often to go for layered defences where a unified combat command capability is provided by an airborne warning and control system (AWACS) or ship-borne system.[34] However defensive saturation is still unlikely to be achieved as it would be difficult to avoid fratricide even if unlimited quantity of interceptors were available.[35] So, despite the arrival of layered missile defence system the defence dominance is held at the operational level by missile systems that are very capable theatre-based deterrents.

This defence dominance has in fact gained in strength with the development of new kind of systems, particularly hypersonic missiles which can strike at long ranges. These non-ballistic hypersonic missiles that move with a speed of Mach 5 or greater and are highly manoeuvrable, accurate, and long-ranged add to the strategic complexity.[36] Their unique characteristics make them effective both against "time sensitive targets and as an anti-access-area denial asset".[37] Hypersonic missiles could be in the form of hypersonic glide vehicles (HGVs), also known as boost-glide vehicles, and hypersonic cruise missiles (HCMs). The former "are non-propelled missiles that glide towards their target at hypersonic speeds after being launched by a rocket into the uppermost layers of the atmosphere". The latter use "traditional booster rockets to bring the projectile up to supersonic speed but then switch to a scramjet propulsion system that enables the missile to sustain hypersonic speed". Ballistic missiles that often travel faster are in fact much faster than hypersonic missiles, but they follow a predictable projectile path. Hypersonic missiles, on the other hand, follow an unpredictable erratic path. Missile defence systems that can estimate the re-entry trajectory therefore are likely to be or efficient against ballistic projectiles but struggle against hypersonics, thus rendering them obsolete. Countries like Russia, for instance, have developed Avangard HGV that can fly at Mach 20 and reach any Western capital within 15 minutes and Kinzhal that has a range of 2000 km which it can cover at Mach 10 speed.[38] These missiles offer a counter deterrent against potentially effective ballistic missile defence systems in future. But they also offer the option of strengthening the conventional capability against theatre-based missile defence systems and buttress current operational-level defensive military postures like anti-access/area-denial[39] that are followed by many nuclear-armed states.

A second important technology is an outcome of kill webs, drones, and drone swarms. They introduce mass into the battlespace that is the quantitative form factor they bring in that affects offence-defence balance as it exists.[40] Their quantitative technological impact could potentially spread across all levels but particularly at the tactical and the operational levels. Combat dynamics has shown that when force levels increase for both advancing and non-advancing forces, initially when force levels are low, the increase benefits the advancing forces as it improves their ability to achieve breakthrough an unprotected section of the defender's front. However, after sufficient force levels are reached the defender's front gets saturated and this eliminates the opportunity for the advancing forces to exploit a weakness in the front, thus benefitting defence.[41] Drones that can offer the quantitative advantage exploit this scaling effect. UAV systems can complement traditional aerial and land forces where their cheaper cost allows them to be sent to penetrate and perform counter-air-defence missions[42] which can then create a window of opportunity for traditional forms of aerial forces and ground-based military units. Although drones have an impact throughout the conventional spectrum at tactical, operational, and strategic levels.

By their very nature their ability to be used on a large scale and their cheap costs make drones an offence heavy military system that can also serve defensive purposes at the operational level. Their remote nature of operation puts the operator away from harm's risk even if a greater number of operators may be required to operate the system, and they are relatively disposable as, even if a few are lost, their large-scale or swarm usage can still allow for neutralization of the target.[43] Thus these are inherently offensive at the platform level. Although drone systems offer capabilities to both sides those on operational defence and those on operational-level offence, ICS-based strategies take advantage of the fact that higher escalation prospects are triggered by use of certain kind of systems, and they affect the balance of resolve for the two sides asymmetrically. An instrumentalized conventional strategy converts this offensive advantage of a system into defensive dominance at the operational level as it aids deterrence and reinforces the military advantage achieved through peacetime coercion at the tactical level. With respect to drone swarms this is very pertinent because as the size of drone swarms increases their ability to take out targets increases, particularly against individual big platforms and air-defence systems; this may free up traditional aerial resources for more close air support (CAS) operations for ground troops. These drones are used to fulfil a variety of functional capabilities at all levels in the conflict spectrum that include their use to guide kinetic strikes by integrating them with command posts, electronic warfare and electronic counter measures, and surveying and mapping of operational geography.[44] They offer persistent reconnaissance of the battlefield at the tactical and the operational levels, although there are restrictions that can be placed on their operation sometimes due to inclement weather.[45] They also enhance the lethality of traditional heavy artillery and aerial forces by complementing them in operations.[46] All this capability is afforded at a very low price, although some high-level individual drones can be expensive. Turkish experience in Syria and Libya as well as the Nagorno-Karabakh clash has shown that kamikaze-style drones are very effective.[47]

Drones, however, do offer capabilities to actors that might want to initiate offence at the operational level in that they provide the ability to penetrate defensive bubbles as part of the operational-level A2/AD doctrines. At the operational level they offer the capability to inflict damage in the operational rear of the adversary,[48] thus giving a tremendous advantage in terms of being able to hit those areas without using manned assets or missiles with a potent weapon that is somewhere lower on the escalation ladder than missiles and be able to inflict the same quantum of damage that can turn the tide of the conflict. This has been shown in practice during the Turkish offensive against Syrian forces in March 2020 where drones were able to hit targets that Turkish F-16s could not.[49] Effectively this is in line with the fact that UAVs can be used to hit operational rear of the adversary. They can do this better because they can be programmed for one-way missions than a manned fighter jet because the latter have to return to refuel and reload. Drones, on the other hand, can maintain operational tempo which can be critical.[50] They also nullify the effect of terrain as troops in mountains and passes could now be more effectively targeted by these drones negating their defensive advantages.[51] Similarly, drones have also had the effect of blunting platforms like tanks in conflicts[52] which are also essential to defence.

Drone defence capabilities encounter problems in that defensive systems can't be employed on a large scale. Laser and other forms of directed energy weapons, for example, are unlikely to be scaled the same way drones can be, and especially against use of drone swarms at the operational or tactical level no current counter-drone technology might prove sufficient.[53] Drones are obviously not invulnerable as many US and British systems were challenged by Russian electronic warfare.[54] The Krasukha electronic warfare system, for instance, first deployed by Russia in Syria in 2015, jams radar and GPS signals as well as other electronic communications and therefore, at least theoretically, enables a foolproof defence against drones.[55] However, while there are counter-drone doctrines that are premised on a large-scale use of electronic counter measures, it is not known if they can be proportionately scaled as easily as drones can be, and even if that happens it is entirely possible that large-scale autonomous drone swarms might be able to carry a few payloads of their own for electronic warfare that could give them resistance against such measures.[56] In fact this gives drone swarms the ability to conduct combined arms warfare on their own given the variety of payloads they can carry to generate both kinetic and non-kinetic effects in pursuance of their mission and thus heavily shape the operational-level environment. Some studies have argued that drones are unlikely to shift the offence-defence balance alone, though when used with infrastructure and other platforms for operational support, they could become significant force multipliers[57] that are likely to aid offence. Other analysts have, however, noted that the drone capabilities and tactics and drone countermeasures are still evolving, and it's too early to judge the impact on the offence-defence balance.[58] On balance, however, these are still more useful as offensive systems at the tactical level that can serve defensive missions at the operational level particularly because they increase the sensor density in the kill web which makes large-scale manoeuvre and surprise difficult if not impossible.

The effect of cyber operations at the operational level of conventional warfare for advancing and non-advancing forces is even more difficult to quantify or qualify. Within the cyber domain preparing an attack takes time because along with network exploitation network presence requires persistent surveillance and understanding of the operational environment of the network to deliver the effect.[59] In the cyber domain increasing number of software vulnerabilities that both an attacker and defender can discover is likely to benefit the attacker initially as it will increase the probability that it can exploit a breach that the defender is yet to discover and patch.[60] However as both discover vulnerabilities eventually the defender would have discovered and fixed all points of probable attack, thus leaving nothing for the attacker to exploit.[61] For this reason the time available to deliver an effect is finite and the actor conducting cyber operations cannot afford to wait. Offensive cyber capabilities have to be tailored to specific target since each malware has to be custom-built and has to be up to date at the time of conflict.[62] Many higher-level military targets like command and control centres are in fact often air-gapped and require social engineering or exquisite intelligence to gain access.[63] These attacks are mostly counter-force in nature, but there are others who have argued that such operations do not destroy conventional forces and the effects that are achieved are temporary or reversible.[64] Cyber capabilities can potentially create an offensive advantage at the operational level if they can aid kinetic aerial operations by suppressing and interdicting adversary data fusion centres and their multiple tactical data links.[65] In effect it could impair the kill web capability of the adversary and therefore their ability to paint a comprehensive picture of the battlefield. In actual combat, for instance in Ukraine, Russian operations saw very little synchronization or correlation between cyber-attacks and kinetic operations, and they did not yield a joint effect.[66] During the US military operations in Libya in 2011 as well doubts were expressed about the reliability of the cyber operations, particularly where some kinetic effect was to be achieved and ground commanders in fact preferred kinetic strikes to disable or take out the target permanently.[67] At the operational level where there could be multiple critical targets to hit, complexity is likely to increase along with coordination problems; therefore, the probability of successful targeting may be low.[68] Overall the effect of the direct bearing of cyber operations at the operational level remains unclear.

The overall impact of missiles, drones, and cyber operations in effect assures that the "threat" that deterrence may fail at the theatre or operational level ensures conventional conflict is essentially kept limited. There is, however, an overall integration that happens when offence-dominant tactical strategies are synergized with defence-dominant operational level, in essence a kind of offence-defence integration that plays out across conventional spectrum but at different levels of warfare. This is different from offence-defence integration as typically understood where capabilities across defensive and offensive systems are integrated like missile defences and missile systems.[69] Instead, this kind of integration plays out at the level of strategy and doctrine rather than individual systems and synergizes tactical goals with operational objectives. Outcome is often strategies that look for tactical gains that are mutually reinforcing in nature with operational-level military posture. As these are essentially kill webs going after each other, it's very likely that this defence dominance at the operational level will dis-incentivize aggressive large-scale conventional operations because by their very nature kill web-centric multi-domain operations will expand the scale of the targeting effort required and will be inherently escalatory.

Nuclear ambiguity and conventional-nuclear intersection

In the nuclear realm over time quantitative expansion of the nuclear arsenals has ensured that offensive strategies like the ability of an attacker to reduce the defender's "retaliatory capability to below a certain assured destruction capability" have become less feasible.[70] This would require a tremendous increase in the ratio of attacker's forces to defender's forces and is unlikely to be the case.[71] Thus offence-defence balance inevitably shifts in favour of defence as the size of the arsenals increases.[72] This, however, has important repercussions downstream for conventional levels of warfare. Nuclear armed states make use of the defence favouring offence-defence balance in the nuclear realm by increasing the conventional-nuclear entanglement or nuclear ambiguity, thus increasing the risk of nuclear escalation. This is particularly an important trait of an instrumentalized conventional strategy that seeks to utilize defence-dominant offence-defence balance in the nuclear realm in a cross-level strategy to strengthen its conventional defence at the operational level in the conventional realm. This is not technically nuclear coercion as there is no advertent action or threat to use nuclear weapons; rather, advertent actions are taken to expand the conventional-nuclear intersection that induces caution at the lower levels of the conventional conflict spectrum, over escalation prospects or a single detonation such that it supports conventional operations within acceptable or preferred thresholds and protects conventional capability. The threat is not that an aggressive nuclear war would transpire to the finish, but that conventional-nuclear intersection could lead to a nuclear precipitation event that may then escalate given the presence of an existing military crisis. This kind of threat exploits natural defence dominance at the nuclear level of the conflict to induce caution in conventional operations, thereby allowing space for escalation control and intra-war deterrence, in essence enhancing the defence dominance at the operational level of the conventional conflict.

Since conventional-nuclear intersection of the conflict spectrum sits at the heart of such a strategy, nuclear armed states derive this intersection from a variety of sources and strategies out of their own volition as well as because of a broader trend in weapons technology. Hersman and Simon point to the fact that conventional-nuclear integration could transpire due to commingling of conventional and nuclear payloads on non-ballistic missile systems, dual-use nature of situational awareness capabilities, and the possibility that escalation may occur rapidly in a non-linear way.[73]

Under instrumentalized conventional strategy the initiator state controls various variables (iteration, scale, duration, etc.) to very meticulously control the progression of a crisis from tactical to operational to strategic and from fait accompli to intra-war deterrence or escalation control respectively. In addition, the initiator instrumentalizes defensive balance at the nuclear level in the conflict spectrum for optimum conventional deterrence at the operational level for intra-war deterrence or escalation control. This ensures that nuclear arsenal serves a purpose beyond the task of deterring existential threats. Any form of conventional-nuclear integration therefore ties into the overall strategy of instrumentalized conventional strategy.

Dual-use weapons systems offer an easy prospect of sourcing conventional-nuclear integration. Particularly the rise of dual-capable missiles both cruise and ballistic that can carry both conventional and nuclear warheads gives rise to nuclear ambiguity.[74] Hypersonic missiles, for instance, can be fitted with nuclear or conventional warheads; they naturally contribute to expanding the conventional-nuclear intersection in the conflict spectrum. The entanglement of conventional and nuclear systems apart from using the same delivery systems for conventional and nuclear roles could also happen due to placing them in the same geographic area or placing them under the command and control of the same entity and using similar employment practices.[75] This allows conventional-nuclear intersection through target ambiguity. Missiles thus give operational-level defence an advantage but crucially introduce nuclear ambiguity both as a weapon of use and as a target in conventional conflict, and therefore allow probable inadvertent paths to nuclear escalation.

Another source of conventional-nuclear intersection are highly accurate conventional weapon systems that can mirror the impact of nuclear weapons. These are commonly referred to as conventional prompt global strikes (CPGS) that can potentially be used to fulfil certain nuclear missions. The natural progression of the introduction of the "triad of nuclear and conventional weapons; conventionally armed ballistic missiles, and the concept of prompt global strike" seems to suggest that "advanced conventional capabilities can well and truly substitute for a spectrum of missions that could previously only be assigned to nuclear weapons".[76] The nature of this conventional-nuclear intersection through CPGS, however, does have other issues to contend with. Critics hold that this conventional set of weapons has undergone a "nuclearization" in the third nuclear age similar to what some early strategists were accused of when they engaged in "conventionalization" that is treating nuclear weapons as just more potent weapons for battlefield effect.[77] The issue of conventional versus nuclear is therefore not a straightforward one as though these non-nuclear weapons are very capable of inflicting critical damage, but it is possible that their deterrent effect may lag their use, as, unlike nuclear weapons, their devastating potential is not universally understood.[78] The perception of the damage inflicted by conventional weapons is also subjective and dependent on the "strategic culture", military, or leadership view.[79] Some may be able to stomach more conventional damage than others, particularly during the escalation-control stage of instrumentalized conventional strategy; conversely some might have very little tolerance for such damage.[80] Analysts have also argued that retaliation by conventional weapons may not have the decisive impact that nuclear weapons could have.[81]

In addition, there are other issues with conventional prompt global strikes which relate to their feasibility and utility. Long-range conventional missiles, though a threat, depend on one central issue in targeting which is actionable intelligence. Precision with regard to the location is important as even a big conventional warhead has relatively small radius of lethality.[82] If there are civilian habitations nearby then they put a limit on the warhead size. A lot depends on remote sensing capabilities. When targeting individuals or single vehicles, imagery and radar satellites are of limited utility as they are not persistent. Where satellites are used to detect a target, detection is likely to be episodic and not persistent.[83] "The geometry of the satellites is constrained by orbital parameters to suboptimal separation of receivers".[84] Combined with the other errors in timing, receiver location, and synchronization, this limits the ability of satellite receivers to precisely locate the emitter. UAVs do fill this gap "to gain intelligence, as they can carry full-motion video, signals intelligence, and other sensors".[85] However, if UAVs can be used to locate near-term targets in future they can also be used to effectively target them, thereby reducing the need for CPGS.[86] And while this may be sufficient for some activities, "it is unlikely to provide the information needed to target conventional ballistic missiles".[87] For near-terms targets this may be insufficient as "a location error of even 1 kilometre would be unacceptable, unless the conventional ballistic missiles (CBMs) have an effective means of terminal guidance".[88] CBMs can effectively target weapons of mass destruction (WMD) launchers and long-range missiles that are difficult to hide, are immobile, and are liquid-fuelled as they will need to be fuelled before launch.[89] However, most such missiles that are readied will give sufficient indication beforehand in a crisis to allow for other conventional means to be deployed more effectively, including airpower. Where sufficient warning is not available even with satellite imagery, CPGS will require actionable intelligence from other sources.[90]

These nuances, however, do not imply that CPGS will not give rise to "nuclear ambiguity" or enlarge the space for conventional-nuclear intersection in the conflict realm. Even if a certain delivery system is identified for CPGS role with a designated conventional warhead, the range and target might ensure that this could still evoke a nuclear response by the adversary. A long-range conventional ballistic missile that may target nuclear infrastructure and may potentially degrade the nuclear deterrent could thus justify a nuclear response.[91] Large-fixed ground targets like bases and other ground or sea-based formations in a given theatre are yet easier to target with conventional missiles pre-emptively in a conventional conflict and still provoke a nuclear response if that base is critical to sustaining operations. China, for example, has considerably invested in fielding hundreds of conventionally armed short-range ballistic missiles (SRBMs) and medium-range ballistic missile (MRBMs) like DF-21.[92] It has deployed cruise missiles capable of carrying conventional and nuclear weapons to ranges of 600 to 2,200 kilometres.[93] Particularly for nuclear weapons states in dyads where they are relatively conventionally weaker to use other conventional means in a hostile military environment, dependency on missiles is likely to be as much of a need as much it may be for show in deterring a more conventionally capable nuclear-armed state. But it could still risk an escalation unlike delivery of conventional munitions by known conventional platforms and weaponry for battlefield effect as missiles will generate both target and warhead ambiguity for the aggressor and the defender. Kill webs can combine with CPGS to make this conventional-nuclear intersection more feasible over time, by virtue of advanced sensor technologies that coupled with unmanned platforms, high-bandwidth networks, quantum computing, data fusion, and artificial intelligence can operationalize conventional lethality to the extent that it may be possible to degrade secure second-strike capability of nuclear-armed states.[94,95] This is because with kill webs there is the future prospect of more accuracy, precession, and timeliness in intelligence. Overall CPGS contribution to conventional-nuclear intersection space is likely to be contextual.

A fundamental source of conventional-nuclear intersection space in the conflict spectrum resides in the nuclear command, control, and communications (NC3) architecture of nuclear-armed states that depends on dual-use space-based assets, meaning they essentially also provide services for conventional operations.[96] Non-nuclear attacks, whether kinetic or non-kinetic, temporary or permanent that can target critical space domain infrastructure, therefore potentially directly threaten nuclear capabilities and evoke nuclear responses and provoke drastic escalation.[97] Even if the objective of degrading space-based capabilities is effectively directed at conventional operations entanglement or spectrum intersection paints any counterspace military effort as a strategic attack that can lead to inadvertent escalation.[98] These capabilities include early-warning systems, including those airborne, communications, intelligence, surveillance, and reconnaissance (ISR); ground-based radars and transmitters; and communication aircraft.[99] These systems are typically dual-use and can be targeted with non-nuclear or conventional capabilities.[100] Sometimes they can even be targeted for purely conventional battles. In fact during the Gulf wars and the counter-terrorism efforts in Afghanistan and Iraq, many strategic situational awareness systems were utilized for conventional operations.[101] This in facts cuts both ways in the emerging setup where even conventional situational awareness systems are now being utilized for strategic warning.[102] States, for example, may target missile defences even if they expect them to be used in conventional contingencies where the theatre conflict is spatially limited.[103] Or it may target its early-warning systems like over the horizon radars other than line of sight radars as these could be used to locate big platforms like aircraft carriers or air breathing systems in a conventional war.[104] In case of cyber and electronic operations which can also create reversible or temporary effects, they add more complexity. These reversible effects present a unique problem in the conventional-nuclear spectrum because they allow hits against other domains particularly space which are very essential to kill webs in the conventional spectrum but are also often dual natured in that they also support nuclear operations. Non-kinetic means like cyber and electronic are more feasible than kinetic means as they allow signal interference with satellites, jamming, spoofing, or attacking the ground segment using cyber or electronic means.[105] These kinds of attacks that don't generate debris allow themselves to be dressed as "conventional" but in terms of the impact they may lead to entanglement of conventional-nuclear spectrums given that most command and control systems are somewhat dual-use in nature as they support both conventional and nuclear operations. These kinds of attacks can sometimes be difficult to detect or attribute as it is difficult to distinguish between a malfunction and an intentional attack.[106] As satellites pass data through ground stations, attacking the ground stations is equally effective in rendering these satellites ineffective.[107] In fact this network that has landlines that link the ground stations to terrestrial networks, user terminals linking satellites and antennas on satellites and ground stations offers multiple points on intrusion and interdiction for non-kinetic means like cyber to potentially disrupt the kill web of the adversary.[108] While electronic warfare attacks like radiofrequency interference can disrupt satellite services, cyber-attacks are more direct and therefore escalatory in nature as they can gain command of the satellite and control its operation.[109] Such attacks can be used to prepare the battlefield for other forms of kinetic conventional operations at both the tactical and operational levels of the war. A certain category of satellites, for instance, could be attacked for their electronic intelligence capabilities that are pivotal to collecting intelligence on military operations.[110] Sometimes their ability to detect the launch of conventional ballistic missiles and provide targeting data that could prove useful to hunt mobile launchers could make them potential targets in conventional operations.[111]

When these non-nuclear means are used, a case may arise where if the targeted state perceives that "non-nuclear attacks against its dual-use C3I assets are preparations for nuclear use when in fact they were motivated by conventional warfighting goals", it could in fact evoke a nuclear response.[112] In most cases countries depend on dual-use C3I architecture for two reasons: one, convenience and savings in cost which would increase if states attempted segregation. Second, segregation could make targeting conventional C3I which is essentially the backbone for operating the kill web easier and more probable given the risk of escalation to nuclear level will be reduced for conventionally stronger nuclear-armed states.[113] In a kill web–centric conventional operational environment the objective and the military doctrine of all adversaries are likely to focus on degrading each other's kill web capacity. In a way conventional-nuclear intersection offers nuclear ambiguity which is instrumentalized to defend this kill web and strengthen gaps in conventional deterrence at the operational level. The flipside is that even if there is segregation in this architecture states may still resort to means which lie in conventional-nuclear intersection space in the conflict spectrum. The pressure to escalate may bear on the targeted state even if the degradation is limited to conventional operations because entire kill webs could be compromised with hits to some nodal systems, pressuring states to use theatre nuclear weapons to convey that redlines have been crossed or to achieve a ceasefire in conventional operations at the level of the theatre in consideration.

There are other strategy interplay-related effects like first-strike instability that contribute to conventional-nuclear intersection and lead to its involuntary expansion. Missile arsenals, for instance, typically require intricate command control networks that are susceptible to damage in a first strike; their presence and potential of use therefore generally increase first-strike instability due to the pressure of use or lose and motivation to go before the adversary does. Missile defence systems add to this pressure as even when they are inefficient or inaccurate, they allow states to adopt first-strike strategies that have the objective of taking out maximum adversary arsenal while neutralizing the remaining arsenal that could be launched at them with imperfect missile defence systems. This is slightly paradoxical in that missile defence systems are actually more supportive of aggressive operations while missiles support defensive operations at the operational level. With hypersonics this first-strike stability issue is compounded as they add another dimension. Hypersonics, with their ability to hit targets of critical value within a short period of time, compress the decision-making time frames to determine whether the projectile is nuclear or conventional and the response needed.[114] This naturally pressures all states to move towards a first-strike capability as defence is at a disadvantage. Nuclear-armed states dealing with the conundrum of defence against missiles have historically tended to favour a trend of preparing for a preventive war as a solution that is achievable at an appropriate cost. This stems from the notion that in a total war the ability to hit first and decisively is a tremendous advantage.[115] As the missile arsenals grew in variety over time with varying level of ranges, warhead capacity, and speed, concepts like "massive retaliation" came into existence that were "like preventive war except that they waited for a provocation".[116] Despite the evolution in these concepts and doctrines, the pressure to launch before sustaining damage has increased due to the presence of missile defence systems and potentially even more effective missile defence systems in future along with cyber vulnerabilities that could compromise the effectiveness of the left over arsenal for a second strike. In 2021 the Chinese in fact tested a hypersonic glide vehicle that circled the globe in low-orbit space before speeding towards its target[117] and seemed to hint towards acquiring capabilities like that of a fractional orbital bombardment system (FOBS)[118] to counter potential development of the space-based sensor layers that might be able to intercept hypersonic missile systems with ground-based interceptors by providing early warning and target quality tracking data.[119] Although these developments are more situated in the realm of nuclear strategy, they offer yet another route to escalation even in a conventional conflict, especially in case of the nuclear-armed state that might have a damage-limitation strategy in the nuclear realm. James Acton writes that states that have offensive nuclear strategies based on damage-limitation doctrine would be dependent on dual-use C3I systems to conduct missile defence operations.[120] If they feel threatened about having reliable C3I during an intra-war deterrence stage or escalation control before the conventional conflict turns nuclear, then there will be an incentive to go first. This can happen regardless of the intent of the adversary, whether it is to degrade conventional or nuclear capability and the means at play kinetic or non-kinetic, thus degrading first-strike stability with the ramification of pushing a conventional conflict into the nuclear realm abruptly.[121] The consequences of expanded conventional-nuclear intersection in the conflict spectrum are a factor that cannot be controlled simply by strategy because multiple factors, including technological trends, might shape this space relative to the bracket of mutual hurt. Even though instrumentalized conventional strategy seeks to exercise escalation control at the operational level and seeks to utilize the expanded conventional-nuclear intersection, this dangerous flirtation does raise the potential for inadvertent escalation spiral between adversaries. The sequential aspect, comunication of cost-benefit form of deterrence, and defence dominance at the operational level attempt to mitigate that risk.

Conclusion

Overall, a large conventional-nuclear intersection or integration within the conflict spectrum through dual-use technology and systems or due to doctrine and strategy related incentives and pressures makes conventional-nuclear intersection a strong feature of the third nuclear age. This invariably dominates conventional operations and operationalization of conventional deterrence as they induce caution in adversaries and forces them to play by the rules of instrumentalized conventional strategies where operational-level environments is shaped over time with tactically offensive strategies and nuclear ambiguities are introduced. These tactically offensive strategies are sequential as brought out in the first chapter and in essence instrumentalize defence-dominant operational-level strategies and in turn defence-dominant nuclear strategies, towards tactical military objectives that yield operational-level outcomes over time.

The mix of missiles conventional, cruise, hypersonics, and missile defences in addition to capabilities like cyber, electronic warfare, and drones that make up concepts like kill web eventually feed into a cross-level instrumentalized conventional strategy where nuclear ambiguity is instrumentalized towards buttressing operational-level conventional capability and deterrence, and the defence dominance at the operational level serves to increase the offence dominance at the tactical level which is used for peacetime coercion. This is an offence-defence integration at different levels of the conflict rather than systems. This allows room even for iterative crises where the conflict envelope is pushed in each iteration with established thresholds coming under pressure over multiple iterations unless there is enough spacing between crises which is essentially contextual to the dyad. These iterative crises, depending on their scale, could technically bifurcate the conventional conflict spectrum into two separate kinds of crises where one kind has a lower scale but are intermittent and cyclical, and second, where the scale is larger and nuclear ambiguity is instrumentalized to limit the scale of the conflict at the operational level, thus establishing a key feature of the third nuclear age where small-scale crises are frequent and appear with a nuclear overhang and potential for escalation.

Notes

1 Robert Jervis (1978) Cooperation under the security dilemma, World Politics, 30:2
2 Charles L. Glaser & Chairn Kaufmann (1998) What is the offense-defense balance and how can we measure it? International Security, 22:4, 44–82, pg. 52
3 Robert Jervis (1978) Cooperation under the security dilemma, World Politics 30:2, 187–194, 199–206
4 Ibid pg. 53
5 Ibid pg. 53
6 Ibid
7 Ibid pg. 73
8 Ibid pg. 55
9 Ben Garfinkel & Allan Dafoe (2019), How does the offense-defense balance scale?, Journal of Strategic Studies, 42:6, 736-763, DOI: 10.1080/01402390.2019.1631810, 758–759
10 Footnoted in Charles L. Glaser & Chairn Kaufmann (1998) What is the offense-defense balance and how can we measure it? International Security, 22:4, 44–82, pg. 53, 54

11 Gerson, Michael, "Conventional Deterrence in the Second Nuclear Age", 2009, https://mca-marines.org/wp-content/uploads/Conventional-Deterrence-in-the-2nd-Nuclear-Age-by-Gerson-090901.pdf. pg. 38
12 Ibid pg .38
13 Ibid pg. 39–40
14 Joy Mitra, "Taking Stock a Year after Balakot: India's Crisis Response and Takeaways", South Asian Voices, March 10, 2020, https://southasianvoices.org/taking-stock-a-year-after-balakot-indias-crisis-response-and-takeaways/
15 Dinakar Peri, Suhasini Haidar & Ananth Krishnan, "Indian Army Says 20 Soldiers Killed in Clash with Chinese Troops in the Galwan Area", The Hindu, June 16, 2020, https://www.thehindu.com/news/national/indian-army-says-20-soldiers-killed-in-clash-with-chinese-troops-in-the-galwan-area/article31845662.ece
16 Dinakar Peri, "A Year after Galwan Clash, China Beefing up Positions along LAC", The Hindu, June 13, 2021, https://www.thehindu.com/news/national/a-year-after-galwan-clash-china-beefing-up-positions-along-lac/article34806645.ece
17 Charles L. Glaser & Chairn Kaufmann (1998) What is the offense-defense balance and how can we measure it? International Security, 22:4, 44–82, pg. 61
18 Ibid pg. 61
19 Ibid pg. 61
20 Ibid pg. 61
21 Ibid pg. 64
22 Ibid pg. 64
23 Ibid pg. 64
24 Footnoted in Charles L. Glaser & Chairn Kaufmann (1998) What is the offense-defense balance and how can we measure it? International Security, 22:4, 44–82, pg. 64
25 This includes the concept of hypersonics, long-range artillery to swarming drones in the conventional battlespace all of which essentially are able to strike at range. See Parsons Dan, "Ukraine War Validates Pursuing Very Long-Range Artillery Army Says", The Drive, May 13, 2022, https://www.thedrive.com/the-war-zone/ukraine-war-validates-pursuing-very-long-range-artillery-army-says
26 Footnoted in Charles L. Glaser & Chairn Kaufmann (1998) What is the offense-defense balance and how can we measure it? International Security, 22:4, 44–82, pg. 64
27 Parsons Dan, "Ukraine War Validates Pursuing Very Long-Range Artillery Army Says", The Drive, May 13, 2022, https://www.thedrive.com/the-war-zone/ukraine-war-validates-pursuing-very-long-range-artillery-army-says
28 Sethi Manpreeth, "Missile Developments in Southern Asia: A Perspective from India", International Institute of Strategic Studies, June 17, 2021, https://www.iiss.org/blogs/research-paper/2021/06/missile-developments-southern-asia
29 Ibid pg. 5
30 Ben Garfinkel & Allan Dafoe (2019) How does the offense-defense balance scale?, Journal of Strategic Studies, 42:6, 736–763, DOI: 10.1080/01402390.2019.1631810, 756
31 Shinichi Ogawa, "Missile Defense and Deterrence", NIDS Security Reports, No. 3, pp. 24–55, March 2002, http://www.nids.mod.go.jp/english/publication/kiyo/pdf/bulletin_e2001_2.pdf. pg. 52
32 Ibid pg. 52
33 Ibid pg. 52
34 Ibid pg. 52
35 Ben Garfinkel & Allan Dafoe (2019) How does the offense-defense balance scale?, Journal of Strategic Studies, 42:6, 736–763, DOI: 10.1080/01402390.2019.1631810, 757
36 Aarten Sander Ruben, "The Impact of Hypersonic Missiles on Strategic Stability", Militaire Spectator, April 21, 2020, https://www.militairespectator.nl/thema/strategie/artikel/impact-hypersonic-missiles-strategic-stability
37 Ibid
38 Ibid

39 Ibid
40 Garfinkel Ben, "Artificial Intelligence, Foresight, and the Offense-Defense Balance", War on the Rocks, December 19, 2019, https://warontherocks.com/2019/12/artificial-intelligence-foresight-and-the-offense-defense-balance/
41 Ibid
42 Kington Tom, "The Drone Defense Dilemma: How Unmanned Aircraft Are Redrawing Battle Lines", Defense News, February 15, 2021, https://www.defensenews.com/global/europe/2021/02/15/the-drone-defense-dilemma-how-unmanned-aircraft-are-redrawing-battle-lines/
43 Leonid Bershidsky, "Why Drones Have Raised the Odds and Risks of Small Wars", The Print, December 2020, https://theprint.in/opinion/why-drones-have-raised-the-odds-and-risks-of-small-wars/554994/
44 Kania Elsa, "The PLA's Unmanned Aerial Systems New Capabilities for a 'New Era' of Chinese Military Power", n.d., https://www.airuniversity.af.edu/Portals/10/CASI/Books/PLAs_Unmanned_Aerial_Systems.pdf. pg. 14–22
45 Urcosta Ridvan Bari, "The Revolution in Drone Warfare: The Lessons from the Idlib De-Escalation Zone", Air University, August 31, 2020, https://www.airuniversity.af.edu/JEMEAA/Display/Article/2329510/the-revolution-in-drone-warfare-the-lessons-from-the-idlib-de-escalation-zone/
46 Ibid
47 Ibid
48 Ibid
49 Ibid
50 Ibid
51 Col Akshaya Handa, "Drone and Counter-Drone Warfare at Tactical Level", Center for Land Warfare Studies, February 8, 2021, https://www.claws.in/drone-and-counter-drone-warfare-at-tactical-level/
52 Rathore Col US, "Intense Tactical Battle Area", Northlines, May 5, 2022, https://www.thenorthlines.com/intense-tactical-battle-area/
53 "How Big of a Deal Are Drone Swarms?", Mad Scientist Laboratory, August 17, 2020, https://madsciblog.tradoc.army.mil/261-how-big-of-a-deal-are-drone-swarms/
54 Kington Tom, "The Drone Defense Dilemma: How Unmanned Aircraft Are Redrawing Battle Lines", Defense News, February 15, 2021, https://www.defensenews.com/global/europe/2021/02/15/the-drone-defense-dilemma-how-unmanned-aircraft-are-redrawing-battle-lines/
55 Leonid Bershidsky, "Why Drones Have Raised the Odds and Risks of Small Wars", ThePrint, December 2020, https://theprint.in/opinion/why-drones-have-raised-the-odds-and-risks-of-small-wars/554994/
56 "Turkey's Top Defense Body Starts UAV Stand-off Jammer Project", Daily Sabah, March 24, 2021, https://www.dailysabah.com/business/defense/turkeys-top-defense-body-starts-uav-stand-off-jammer-project
57 See Antonio Calcara, Andrea Gilli, Mauro Gilli, Raffaele Marchetti & Ivan Zaccagnini (Spring 2022) Why drones have not revolutionized war, International Security, 46:4, 130–171
58 Analysts like Zachary Kallenborn say that the general class of drones are still evolving and therefore larger theoretical conclusions on impact of drones are premature. Twitter, 13 May 2022, https://twitter.com/ZKallenborn/status/1524868870045159424?s=20&t=vY121Nm1OzxhQ1copYZLEw
59 Kartik Bommakanti, "Observer Research Foundation", ORF, June 18, 2021, https://www.orfonline.org/expert-speak/india-and-cyberspace-balance-between-offence-and-defence/
60 Garfinkel Ben, "Artificial Intelligence, Foresight, and the Offense-Defense Balance", War on the Rocks, December 19, 2019, https://warontherocks.com/2019/12/artificial-intelligence-foresight-and-the-offense-defense-balance/

61 Ibid
62 Schulze Matthias, "Cyber in War: Assessment of the Strategic, Tactical and Operational Utility in Military Cyber Operations", 12th International Conference on Cyber Conflict, May 2020, https://ccdcoe.org/uploads/2020/05/CyCon_2020_10_Schulze.pdf. pg. 187
63 Ibid pg. 187
64 Ibid pg. 187
65 Bonner III E. Lincoln, "Cyber Power in 21st-Century Joint Warfare", Joint Force Quarterly 74, National Defense University Press, 1 July 2014, https://ndupress.ndu.edu/JFQ/Joint-Force-Quarterly-74/Article/577536/cyber-power-in-21st-century-joint-warfare/
66 Schulze Matthias, "Cyber in War: Assessment of the Strategic, Tactical and Operational Utility in Military Cyber Operations", 12th International Conference on Cyber Conflict, May 2020, https://ccdcoe.org/uploads/2020/05/CyCon_2020_10_Schulze.pdf. pg. 190
67 Ibid pg. 190
68 Ibid pg. 194
69 Green Brian, "Offense–Defense Integration for Missile Defeat", Missile Threat, July 6, 2020, https://missilethreat.csis.org/offense-defense-integration-for-missile-defeat/
70 Charles L. Glaser & Chairn Kaufmann (1998) What is the offense-defense balance and how can we measure it?, International Security, 22:4, 44–82, pg. 52, 53, 58, 66
71 Ibid pg. 56, 58, 66
72 Ibid pg. 66
73 Rebecca K.C. Hersman & Maxwell Simon, "Examining International Security at the Nuclear Nexus", Defense News, February 8, 2021, https://www.defensenews.com/opinion/commentary/2021/02/08/examining-international-security-at-the-nuclear-nexus/
74 "China's Growing Missile Arsenal and the Risk of a "Taiwan Missile Crisis" , Nuclear Threat Initiative, 2020, https://www.nti.org/analysis/articles/chinas-growing-missile-arsenal-and-the-risk-of-a-taiwan-missile-crisis/
75 David C. Logan (2020) Are they reading Schelling in Beijing? The dimensions, drivers, and risks of nuclear-conventional entanglement in China, Journal of Strategic Studies, 2, https://www.tandfonline.com/doi/full/10.1080/01402390.2020.1844671
76 Gerson Michael, "Conventional Deterrence in the Second Nuclear Age", 2009, https://mca-marines.org/wp-content/uploads/Conventional-Deterrence-in-the-2nd-Nuclear-Age-by-Gerson-090901.pdf
77 Ibid pg. 33
78 Shinichi Ogawa, "Missile Defense and Deterrence", NIDS Security Reports, No. 3, pp. 24–55, March 2002, http://www.nids.mod.go.jp/english/publication/kiyo/pdf/bulletin_e2001_2.pdf. pg. 31
79 Ibid pg. 31
80 Ibid pg. 31
81 Ibid pg. 31
82 Long, Austin, Dinshaw Mistry & Bruce M. Sugden (2010) Going nowhere fast: Assessing concerns about long-range conventional ballistic missiles, International Security, 34:4, 166–184. Accessed May 26, 2021, http://www.jstor.org/stable/40784565. pg. 166
83 Ibid pg. 166
84 Ibid pg. 166
85 Ibid pg. 166
86 Ibid pg. 166
87 Ibid pg. 166
88 Ibid pg. 166
89 Ibid pg. 166

90 Ibid pg. 166
91 Ibid pg. 176
92 ibid pg. 181
93 ibid pg. 181
94 Cited in Rebecca Hersman & Reja Younis, "Surveillance, Situational Awareness, and Warning at the Conventional-Strategic Interface", Center for Strategic and International Studies, January 15, 2021, last modified January 15, 2021, https://defense360.csis.org/surveillance-situational-awareness-and-warning-at-the-conventional-strategic-interface/. Pg. 5
95 Ibid pg. 8
96 Ibid
97 Ankit Panda, "Space-Based Nuclear Command and Control and the 'Non-Nuclear Strategic Attack'", The Diplomat, April 8, 2020, https://thediplomat.com/2020/04/space-based-nuclear-command-and-control-and-the-non-nuclear-strategic-attack/
98 Ibid
99 James M. Acton (2018) Escalation through entanglement: How the vulnerability of command-and-control systems raises the risks of an inadvertent nuclear war, International Security, 43:1, 56–99, pg. 57
100 Ibid pg. 57
101 Hersman Rebecca (2020) Wormhole escalation in the new nuclear age, Texas National Security Review, 3, 90–109, pg. 4
102 Ibid
103 Cited in James M. Acton (2018) Escalation through entanglement: How the vulnerability of command-and-control systems raises the risks of an inadvertent nuclear war, International Security, 43:1, 56–99, pg. 66
104 Ibid pg. 78
105 Kartik Bommakanti, "India and cyberspace: Balance between offence and defence", ORF, June 18, 2021, https://www.orfonline.org/expert-speak/india-and-cyberspace-balance-between-offence-and-defence/
106 Rajagopalan Rajeswari, "Electronic and Cyber Warfare in Outer Space", UNIDIR Space Dossier 3, May 2019, https://www.unidir.org/files/publications/pdfs/electronic-and-cyber-warfare-in-outer-space-en-784.pdf. pg. 3
107 Ibid. pg. 9
108 Ibid pg. 9
109 Ibid pg. 11
110 James M. Acton (2018) Escalation through entanglement: How the vulnerability of command-and-control systems raises the risks of an inadvertent nuclear war, International Security, 43:1, 56–99, pg. 85
111 Ibid pg. 85
112 Ibid pg. 67
113 Zhao, T., & Bin, L. (2017). The underappreciated risks of entanglement: A Chinese perspective. Entanglement: Russian and Chinese Perspectives on Non-Nuclear Weapons and Nuclear Risks, pg. 68
114 Aarten Sander Ruben, "The Impact of Hypersonic Missiles on Strategic Stability", Militaire Spectator, April 21, 2020, https://www.militairespectator.nl/thema/strategie/artikel/impact-hypersonic-missiles-strategic-stability
115 Brodie Bernard. Strategy in the Missile Age. Princeton, NJ: Princeton University Press, 2015.
116 Ibid
117 Sevastopulo Demetri & Kathrin Hille, "China Tests New Space Capability with Hypersonic Missile", Financial Times, October 17, 2021, https://www.ft.com/content/ba0a3cde-719b-4040-93cb-a486e1f843fb
118 "This was developed by the Soviet Union in the 1960s and was operationally deployed in the 1970s to bring a missile into Earth's low orbit. Such a trajectory

would allow the missile system to come from the south via Antarctica and evade the Arctic, where U.S. early-warning radar detection is concentrated." See Bugos Shannon, "China Tested Hypersonic Capability, U.S. Says", Arms Control Association, November 2021, https://www.armscontrol.org/act/2021-11/news/china-tested-hypersonic-capability-us-says

119 Kelley M. Sayler & Stephen M. McCall, "Hypersonic Missile Defense: Issues for Congress", Congressional Research Office, January 26, 2022, https://www.everycrsreport.com/files/2022-01-26_IF11623_fd5452651d4f2bb38f06885e6e903cf04830d100.pdf

120 This, however, would be a violation of instrumentalized conventional strategy that should ideally seek to control escalation at the operational level of the conventional war.

121 James M. Acton (2018) Escalation through entanglement: How the vulnerability of command-and-control systems raises the risks of an inadvertent nuclear war, International Security, 43:1, 56–99, pg. 59

CONCLUSION

Instrumentalized conventional strategy is a concept that essentially seeks to establish linkages between different domains, technologies, and military strategy in a conflict spectrum which is essentially a conventional-nuclear continuum rather than a conflict spectrum which is divided into discrete levels of conventional and nuclear warfighting space. Strategy and technology both have contributed to shaping the conflict spectrum this way, and so the topic of conventional deterrence or nuclear deterrence demands to be treated and tied within a unified concept and requires unified military doctrine. Conventional deterrence is still pivotal, but the nature of the conflict spectrum and fait accompli–based military strategies introduce the dimension of sequence in deterrence and strategy. Sequential strategies inherently allow for two things: first, crisis progression and, second, merging of different levels of the conflict: tactical, operational, and strategic which correspond to peacetime coercion, intra-war deterrence, or escalation control, and conventional-nuclear intersection respectively in the conventional strategy space. Peacetime coercion at the tactical level is a compellent activity, and offence dominance is enjoyed through a strategy of calibrated gains which requires a thorough understanding of adversary thresholds. Conventional forces and conventional military posture at the operational level are involved and support this activity to create new military ground situations and establish gradual changes that can have operational-level impact over time. These calibrated military gains are essentially mutually reinforcing with newly acquired operational-level military posture and disallow the adversary to regroup and mount a counterattack at the tactical level rather than essentially demanding an operational-level counteroffensive which, due to its more ambitious military goal, effort, and resources, inherently is pitted against the defence dominance enjoyed at the operational level.

At the operational level of the conflict, various existing, evolving, and emerging technologies and concepts driven by those technologies contribute to the defence dominance and allow for intra-war deterrence and escalation control to be practised at the operational level. At this operational level the operationalization of cross-domain sensing and shooting could allow kill chains to evolve into more lethal kill webs that create a common operating picture with platforms based on land, sea, and air and combine capabilities in the cyber, electronic, and space domains. This presents an unprecedented level of situational awareness through a large sensor density in the battlespace that, combined with great accuracy, precision lent through intelligent weapons, missiles conventional, hypersonics and cruise, unmanned vehicles and unmanned autonomous vehicles, allows offensive capability that is very effective. By its very nature the kill web introduces speed warfare that is as effective as it escalatory, but this is moderated by the established military doctrine based on an instrumentalized conventional strategy. Kill web is at the heart of the concept of throwing in operational-level deterrence and capability at tactical-level conflicts, thus erasing the distinction between the operational level and the tactical level and is a key feature of the third nuclear age. This kill web is supplied with mass by the introduction of drones that could potentially be more escalatory when they are used at the tactical or the operational level both as a means and as a target, as they incentivize kinetic action. These capabilities are brought into a single and joint all-domain command and control that allows kill web to support defensive objectives at the operational level, though they also afford offensive capability to conduct missions in operational rear areas. This kind of command and control is likely to face a trade-off between having a more effective, highly autonomous but potentially escalatory system of systems versus having a system of systems that are less effective, autonomous with a human in the loop with potentially greater control over escalation and progression of crises.

Instrumentalized conventional strategy instrumentalizes conventional capability at the operational level towards peacetime coercion at the tactical level, but the sequential nature does not end here and it also instrumentalizes nuclear ambiguity towards operational-level conventional deterrence. Essentially this has the ramification that hitherto nuclear deterrence served only existential threats but now it has role to address even threats in the conventional spectrum by limiting the potential of conventional capability at the operational level available to the adversary. This is largely due to expansion in the conventional-nuclear conflict spectrum intersection on counts of, first, ambiguity related to targets and, second, ambiguity related to means. Dual-use missiles, weapon platforms, co-location, or co-mingling of conventional and nuclear arsenals and, above all, dual-use nature of situational awareness capabilities that serve nuclear as well as conventional operations contribute to this entanglement or conventional-nuclear intersection. Non-kinetic capabilities like cyber and electronic could crucially threaten the dual-use situational awareness capabilities with reversible or permanent effects, effectively pushing the conflict in the conventional-nuclear intersection. There are other trajectories, some by design and some of which are not by design but provoked by a strategy like the generation of first-strike stability issues, that could allow for a conventional-nuclear transition in the conflict spectrum. These are the use of tactical nuclear weapons as a means of attaining status quo in a conventional conflict or their use for demonstrative purposes when critical military interests are at stake. Use of long-range conventional missiles for hitting critical leadership or nuclear infrastructure targets that might provoke a nuclear response. Drones and kill web also provide the capability to make nuclear second-strike capability vulnerable especially in the undersea domain which is considered relatively invulnerable, and this could generate a route for nuclear escalation due to the "use or lose" dilemma. In the nuclear realm if a state has a "damage-limitation" strategy, then even in a conventional war if its early warning and situational awareness capabilities are threatened, it could provoke a nuclear strike while its nuclear infrastructure is operational and dual-use C3I capabilities are intact. That states are willing to use nuclear ambiguity or escalation prospects towards escalation control also allows space for misperception, miscalculation, and escalation spiral and is a key concern around the third nuclear age. Instrumentalized conventional strategy essentially seeks to control such crises by ensuring intra-war deterrence, iterative conflict, and gradual operational-level gains attempt to balance the trade-offs involved in the use of new technologies and the entire conflict spectrum to guide the progression from tactical to operational in a calibrated but deliberate manner.

Due to the sequence aspect, there is a lot of thrust at the local-level military equation in this kind of strategy. Overall deterrence can only be as effective as local-level deterrence or balance of forces or balance of capabilities. Even with this balance adversary forces will look to exploit the offence dominance in time-space at the local level. This would make local force posture fundamental to maintaining overall deterrence between nuclear-armed states and for the nuclear revolution understood as the absence of direct conflict between nuclear-armed adversaries, as this will introduce a fundamental change with frequent direct and iterative crises. Two kinds of frameworks could be used to assess such strategic stability. One could be the traditional understanding of segregated conflict spectrum with separate deterrence frameworks of conventional and nuclear deterrence, and second, a framework that sees the conflict spectrum as one single entity with the concept of military deterrence or integrated deterrence which could be an area of future research for scholars. This has implications for where arms control efforts should focus and might get complicated for the same reason which is to first untangle the entanglement. This could be a difficult proposition to achieve given, first, if its deliberate strategy by the adversary to instrumentalize nuclear ambiguity; second, even if the adversary is willing costs involved in segregation might not favour such segregation; third, such segregation could allow the more competent conventionally capable nuclear state to target the weaker conventional adversary's conventional kill web. A second important conversation could revolve around the kind autonomy that is bestowed on the joint all-domain command and control for the kill web. The norms around the level and nature of this autonomy are likely to face a trade-off between effectiveness and escalation and, left to individual nuclear-armed states, could be a cause for inadvertent escalation in conventional conflicts even if instrumentalized conventional strategy is operational in the dyad. These impediments could imply that the space for arms control strategies may likely shrink in the future, giving way to more dependence on deterrence for stability unless a major breakthrough is made in this area or a new technology makes it possible to lessen target or warhead ambiguity. Dependence on deterrence is likely to focus on military posture and forward deployment of forces for either deterrence by denial or deterrence by punishment strategies where the latter is geared more towards intra-theatre targets rather than inter-theatre targets for a military response. One way to surmount this challenge could be to devise some sort of theatre or very limited versions of arms control which are contextual and local in nature while efforts are on to achieve more grander versions for arms control at the international level. This is with the understanding that political intent will matter a lot, and technology is being led by strategy, not the other way round as the key to TNA is that ICS-based strategies have essentially come into force because the political intent is offensive.

In essence, therefore, the third nuclear age is a sum of developments in the conventional and nuclear domain in terms of both technology and strategy that shape conventional deterrence and make it pivotal in both lower- and higher-order conventional conflicts. However, emerging technology per se is not as much responsible for the "strategic complexity" that is attributed to it in the third nuclear age, and to a large degree conscious military strategy and related decisions are still primarily responsible for this complexity.

INDEX

For Product Safety Concerns and Information please contact our EU
representative GPSR@taylorandfrancis.com
Taylor & Francis Verlag GmbH, Kaufingerstraße 24, 80331 München, Germany

www.ingramcontent.com/pod-product-compliance
Ingram Content Group UK Ltd.
Pitfield, Milton Keynes, MK11 3LW, UK
UKHW022314250926
475702UK00046B/780

* 9 7 8 1 0 3 2 4 1 9 7 2 5 *